Elimination of micro-organisms by Drinking Water Treatment Processes

KWR Watercycle Research Institute Series

Elimination of micro-organisms by Drinking Water Treatment Processes
A review

by W.A.M. Hijnen and G.J. Medema

IWA Publishing
London • New York

KWR

Published by **IWA Publishing**
Alliance House
12 Caxton Street
London SW1H 0QS, UK
Telephone: +44 (0)20 7654 5500
Fax: +44 (0)20 7654 5555
Email: publications@iwap.co.uk
Web: www.iwapublishing.com

First published 2010

Typeset in India by OKS Prepress Services.

British Library Cataloguing in Publication Data
A CIP catalogue record for this book is available from the British Library

Library of Congress Cataloging-in-Publication Data
A catalog record for this book is available from the Library of Congress

ISBN 10: 1843393735
ISBN 13: 9781843393733

Contents

 Elimination of micro-organisms by Drinking Water Treatment Processes. By W.A.M. Hijnen and Gertjan J. Medema. ISBN: 9781843393733. Published by IWA Publishing, London, UK.

Preface

The Dutch Drinking Water Decree prescribes that drinking water companies should demonstrate that their drinking water is microbiologically safe by performing a quantitative microbial risk assessment (QMRA). The required level of safety is a risk of infection of 10^{-4} per person per year. Such a QMRA requires knowledge about (i) the concentration of pathogens in source water and (ii) the efficacy of water treatment to eliminate these pathogens. Consequently, there is a need for a scientific database with the elimination capacity of all relevant processes used in drinking water treatment for viruses, bacteria and pathogenic protozoa. In the joint research program of the Dutch Drinking water Companies (BTO-programme) a project was defined to create such a database.

An increasing amount of knowledge about the microbiological efficacy of drinking water treatment is published in the literature. This literature is collected and evaluated. Many publications describe experiments on the elimination of micro-organisms that are conducted in the laboratory under well-defined conditions with lab-strains of micro-organisms. Experience of KWR and of many others show that the efficacy of (well-controlled) full-scale treatment processes is usually lower than expected from a direct extrapolation of laboratory experiments. This can be due to several factors, such as the variability of the conditions in practice (feed water quality, temperature etc.), hydraulic differences between well-mixed laboratory vessel and a large flow-through reactor and the difference between micro-organisms in the laboratory and in the environment (survival state, attachment to particles etc.).

This does not disqualify laboratory experiments. These provide a first impression of the efficacy of a process and in some cases data collection of full-scale conditions is not possible. Furthermore, for disinfection processes these studies are necessary to assess dose-response curves for selected pathogenic micro-organisms under standardised conditions. In addition laboratory experiments are needed to study the effect of conditions such as temperature, pH, turbidity etc. on the efficacy of a process. Combining these results with full-scale observations will optimise the process of risk assessment.

The overall aim of this review was to produce a default value for the Micro-organism Elimination or Inactivation Credit (MEC or MIC) of full-scale treatment processes and a description of the effect of water quality parameters and process control parameters on the elimination or inactivation capacity. The literature data are valued according to their representation of full-scale conditions.

- For physical processes the calculated default value of MEC is weighted on the basis of the resemblance with full-scale conditions.
- In the dose requirement table described for certain MIC values of a disinfection process, effects of microbial and process conditions on the dose-response curves assessed for spiked and pre-cultured organisms are included.

Third edition in 2010

The report describes the state-of-the-art of most of the literature from 2006 and before and is supplemented with few later published papers. The literature on conventional treatment (coagulation and floc-removal plus rapid granular filtration) was added to the second edition (Chapters 3–5) and the Chapters on slow sand filtration and UV disinfection were updated.

Acknowledgement

This report has been written as part of the Joint Research Program of the Dutch Water Companies (BTO). The collaboration with the different Water Companies to collect and analyze 'real world' data on the elimination of indicator bacteria by their water treatment processes and to conduct additional challengetests at pilot plant scale (Waternet, Brabantwater, Evides) is highly appreciated.

1 Introduction

Recently the Dutch Drinking Water Decree drinking (WLB, 2001) has been revised. In this new decree, the microbial safety of drinking water is regulated by the traditional parameters (*E. coli* and enterococci 0 per 100 ml) and a new risk-based standard is introduced. Surface water supplies need to demonstrate the safety of their drinking water with a Quantitative Microbial Risk Assessment (QMRA). The maximum risk level is defined (provisional) as a risk of infection of 10^{-4} per person per year (pppy). Such QMRA requires knowledge about (i) the concentration of pathogens in the source water to calculate the required elimination capacity and (ii) about the efficacy of water treatment to eliminate them.

The legislative microbiological monitoring programme is not appropriate and designed to assess the elimination capacity of water treatment processes for pathogenic micro-organisms. Therefore water companies perform a more extended microbiological monitoring programme with parameters such as indicator bacteria (coliforms, enterococci and spores of sulphite-reducing clostridia) and heterotrophic plate counts at selected sampling points. For QMRA the use of these data is either not appropriate or too limited. Heterotrophic plate counts reflect biomass concentration fluctuation due to removal and growth in the processes. Moreover, concentrations of indicator bacteria are usually too low and large volume sampling is needed to assess elimination by the full treatment.

Many scientific publications are available on the effect of disinfectant chemicals and UV on micro-organisms and transport of micro-organisms in filter materials and soil. A few of these describe the removal of pathogenic and indicator bacteria by full treatment processes (Gale *et al.*, 2002; Hijnen *et al.*, 2000; Hijnen *et al.*, 2004; LeChevallier *et al.*, 1991; Medema *et al.*, 2003; Payment *et al.*, 1985; Payment and Franco, 1993).

The aim of this study was to evaluate available literature data and create a well-defined database which enables calculation of the Micro-organism Elimination or Inactivation Credit (MEC or MIC) for viruses, bacteria and oocysts of pathogenic protozoa of all relevant processes used in drinking water treatment. The results of this review will also be used to identify gaps in the current knowledge necessary to identify further research needs.

Chapter 2 describes the data collection and the evaluation methods. Chapters 3, 4, 5, 6 and 7 of this third edition describe the results for conventional treatment (coagulation, floc-removal and rapid filtration), coagulation and primary floc removal, rapid granular filtration, slow sand filtration and UV disinfection, respectively.

2 Data collection and evaluation methods

2.1 Qualification of the reviewed literature data

Literature on the removal of micro-organisms by water treatment processes has been collected and evaluated on the technical and microbiological aspects of the study. From each of these publications the Decimal Elimination Capacity [equivalent to ^{10}log-removal or decimal reduction] was calculated. In the process of literature reviewing the studies were qualified according to two main criteria:

(1) Reliability of the collected data and availability of the process conditions in the study;
(2) Information on the conditions of the studied processes and the deviation from the situation under full-scale water treatment.

The *first* criterion is used for a selection of studies to be used for the calculation. In the review, only those studies were used where inactivation and removal was assessed using generally accepted microbial culturing (solid media or tissues) or animal infectivity methods. The *second* criterion is used to calculate a weighted average DEC or the Micro-organism Elimination Credit (MEC) from all individual studies. Both criteria will be clarified hereafter.

Physical processes. The average DEC of all individual studies is used to calculate the Micro-organism Elimination Credit MEC and the range of DEC (minimum and maximum) observed for the process. These data are presented in the overall matrix of data.

Disinfection processes. For every individual micro-organism the average inactivation rate constant k is presented, calculated from the accumulated data. With this parameter the Micro-organism Inactivation Credit (MIC) is calculated for a range of at least three time-related doses or UV-intensities (concentration and time combination; C∗T or UV-fluence) used under full-scale conditions.

The method of accumulation of all DEC data from the individual studies to one matrix MEC- or MIC-value is described hereafter.

2.1.1 The quality of the study

Studies on *physical removal* are judged on:

– availability and quality of the technical process conditions;
– description and quality of the experimental conditions;

 Elimination of micro-organisms by Drinking Water Treatment Processes. By W.A.M. Hijnen and Gertjan J. Medema. ISBN: 9781843393733. Published by IWA Publishing, London, UK.

- description and quality of the microbiological data;
- quality assurance;
- reproducibility of the experiments.

The evaluation results in a quality score (Table 2.1). If the total number of studies is six or more, the MEC is calculated from studies with an A-score only. When the number of studies is limited (≤5), studies with a B-score are also used.

Table 2.1 Quality score of the study

	Many process/study conditions	Few process/study conditions
Reproducible results	A	B
Variable results	B	C

For review of the *disinfection* processes only studies were evaluated in which assessment of dose or UV-fluence was clearly described and based on either proper assessment of the disinfectant concentrations or UV sensor measurements, fluence calculations and/or Reduction Equivalent Fluence (REF) assessment with biodosimetry. The description and quality of the microbiological data, quality assurance and reproducibility of the experimental data, as well as the availability and quality of the technical and experimental conditions were reviewed. Only those studies where process and experimental conditions were well documented were used.

2.1.2 Physical processes

The DEC-values of the studied processes are weighed based on the following two criteria:

- The scale of the studied process: full-scale systems, pilot-plant study or laboratory experiments;
- The type and conditions of the micro-organisms used: elimination of environmental organisms, lab-cultured organisms or model-organisms (surrogate parameters).

Studies are weighed according to the two-dimensional matrix presented in Table 2.2. Every study is qualified with the Full-Scale index (FS), a parameter used in the final accumulation of the data for the calculation of MEC.

Table 2.2 Score-matrix for attribution of Full-Scale index of studied physical processes

Character of the study	Environmental organisms	Cultured organism	Surrogate parameter
Full-scale processes	5	4	3
Pilot plant			
– demonstration plant	5	4	3
– small scale	4	3	2
Laboratory study			
– continuous flow	4	3	2
– Bench scale	3	2	1

2.1.3 Disinfection processes: UV

Experimental conditions. The inactivation of micro-organisms by UV irradiation has been studied under different experimental conditions. Many studies used a collimated beam apparatus (CB-tests) under bench-scale and well-defined laboratory conditions. A volume of inoculated water is irradiated during varying periods of time under a lamp emitting UV light. Other studies used continuous flow systems (CF-systems) in a laboratory, in a pilot- or demonstration plant or under full-scale conditions where the water passes a reactor with one or more UV lamps and UV irradiation a contact time vary over the reactor.

Drinking and wastewater studies. UV disinfection is applied in the drinking water industry but also for disinfection of treated wastewater. Studies from both applications were reviewed and identified as such. Papers on wastewater studies (WWS) usually describe disinfection of secondary effluent with or without an additional pre-treatment. In the drinking water studies (DWS), water is generally of much lower turbidity and higher UV transmission than in waste water. When relevant, the influence of these parameters is discussed.

UV fluence data. UV fluence cannot be measured directly, so it has to be inferred from monitoring the UV irradiance with a UV sensor and the time that the micro-organisms are exposed to UV. For the Collimated Beam experiments, the average UV irradiance and contact time are well-characterised and have small confidence intervals. For continuous flow systems the average UV fluence can be calculated from the same parameters. However, the confidence intervals are much larger, due to the much larger variation in contact time and in UV irradiation at different points in the reactor, compared to the Collimated Beam experiments. By modelling the hydraulic retention time in the UV reactor using Computational Fluid Dynamics (CFD) accuracy of these calculations has increased the last few years. Alternatively, the Reduction Equivalent Fluence of CF-systems can be determined with biodosimetry (Quals and Johnson, 1983a; Sommer *et al.*, 1999; Õsterreichisches normungsinstitut, 1999; Hoyer, 2004; Pirnie *et al.*, 2003). Biodosimetry is performed by challenging the UV reactor with a micro-organism with calibrated UV inactivation kinetics (biodosimeter) assessed with CB-tests in the test water. With the measured inactivation of the biodosimeter and the calibration curve, the Reduction Equivalent Fluence or REF (mJ/cm^2) can be calculated.

2.2 Calculation of MEC and MIC

2.2.1 *Physical processes*

MEC is the weighted average DEC of all A-studies ($N>5$) or both A and B-studies ($N \leq 5$) and calculated with the following equation:

$$MEC = \frac{\sum(DEC * FS)}{\sum FS} \qquad (1)$$

For every combination of micro-organism and process the strength of the calculated MEC is described by the following two characteristics of the available database:

- the number of DEC-data (N_{data});
- the average FS-index of all studies used to calculate MEC.

In the final matrix where all processes are presented the strength of the MEC is described by Q which is the product of the number of data and the average FS-index.

2.2.2 *Disinfection UV*

The inactivation kinetics of a large number of pathogenic micro-organisms and indicator micro-organisms that are significant for the microbial safety of water have been calculated from studies where UV fluence has been determined under optimal conditions: CB-tests with "drinking water" (low turbidity and high UV transmission value).

Inactivation by UV is based on the damage caused to the nucleic acids (DNA/RNA) of the cell or virus. Primarily the formation of pyrimidine dimers, but also of other photoproducts of nucleic acids and nucleic acid lesions (von Sonntag *et al.*, 2004), inhibit replication and transcription and hence, prevent the cell or virus from multiplying. The highest UV absorbance of DNA is at wave lengths around 260 nm; at lower and higher wavelengths the absorbance decreases. Most studies used low-pressure mercury lamps with a major wavelength output (85%) at 254 nm (monochromatic UV radiation) but for some micro-organisms the UV sensitivity was (also) determined with polychromatic UV radiation from medium pressure lamps.

The UV sensitivity of the selected micro-organisms is described by the parameters of the inactivation kinetics. Inactivation is defined as the reduction of the concentration of culturable micro-organisms N due to the exposure to a concentration disinfectant C during a specific contact time *t*. The inactivation kinetic for chemical disinfectants is most commonly described by the first-order disinfection model of Chick-Watson (1908) and the same model can be applied for UV disinfection. The inactivation of micro-organisms is usually described by the log inactivation of *N*. Based on the first-order model the linear relationship between log inactivation and the UV dose or fluence is described by:

$$^{10}\log\left(\frac{N_t}{N}\right) = -k * Fluence \quad (2)$$

N_t is the microbial concentration after contact time *t*. Fluence is the product of the UV fluence rate (mW/cm^2) and the exposure *t* ($mWs/cm^2 = mJ/cm^2$). In the literature, two main deviations from first-order UV disinfection kinetics have been observed. Some authors (Knudson, 1985; Hoyer, 1998; Sommer *et al.*, 1998; Mamane-Gravetz and Linden, 2005) observed no inactivation of bacteria or bacterial spores at low UV fluences followed by a normal log-linear relationship at higher UV fluences. This is called a shoulder model and is presented by the following equation

$$DI = -k * Fluence - b \quad (3)$$

where *b* is the *y*-intercept, a negative value since the curve is crossing the *x*-axis at the UV fluence where log-linear relationship starts. The second deviation from the log-linear kinetics is a reduced inactivation at high fluences, called tailing. Tailing is not included in the k-values presented in this review, by excluding the inactivation data at higher fluences in studies with tailing (as determined from the plots of inactivation versus fluence).

2.3 Process conditions

2.3.1 *Physical processes*

The influence of the conditions (water quality, process parameters etc.) on the efficacy of treatment processes will not be discussed in detail, but some general remarks will be made and related to the range of values used to calculate the MEC to indicate the major water quality parameters and process conditions affecting the removal efficiency.

2.3.2 *Disinfection UV*

The inactivation kinetics can be used to determine the disinfection efficacy or Micro-organism Inactivation Credit (MIC; log) of full-scale UV systems and to assess the fluence requirement to obtain a certain MIC. However, for translation of CB-results to full-scale UV systems it is essential to know the effect of process conditions on the efficiency of the radiation process. In contrast to oxidative disinfection processes with chemicals like chlorine and ozone the efficacy of UV disinfection is not affected by conditions like temperature, pH (Severin *et al.*, 1983; Pirnie *et al.*, 2003) and reactive organic matter.

UV absorbance by organic and inorganic matter is included in the UV fluence calculation. But the following factors may affect the efficiency of UV disinfection at full-scale:

- Factors related to the micro-organisms: physiological state (pre-culturing, growth phase), strain diversity, repair mechanisms and particle association;
- Factors related to the fluence assessment: fluence-distribution due to the distribution of the hydraulic retention time, adsorption, reflection and refraction of UV light through the water and lamp intensity (aging and fouling).

Several studies have addressed these aspects and are discussed to determine whether adaptation of the MIC or required fluence is required applying the UV sensitivity data to full-scale UV systems and, if possible, to quantify to what extent.

2.4 Presentation of the results

2.4.1 Detailed information

Detailed information about the reviewed literature and the calculation sheets of MEC and MIC are presented in the annexes.

2.4.2 Chapters and paragraphs

Each treatment process is described in a separate chapter subdivided in paragraphs for the elimination of the three main groups, viruses, bacteria (including bacterial spores) and protozoa. References first mentioned are **printed bold;** subsequently these references are printed in regular typeface.

2.4.3 Overall matrix

In the last chapter of the report the overall matrix is given.

3 Conventional treatment

Coagulation/flocculation followed by rapid granular filtration is used world wide in water treatment to produce drinking water from surface water. This process is called conventional treatment. The basic principle behind the process is removal of suspended solids by destabilization of the colloid materials by chemicals, coagulation and primary and secondary flock-removal. Primary flock-removal can be done by sedimentation (SED), lamellae separation (LAM), sludge blanket clarifiers (SBC), dissolved air flotation (DAF), tube-settlers (TUB) and a pulsator (PUL). Rapid filtration over granular media is the secondary floc-removal process. In this chapter the cumulative effect of both processes is presented for the different groups of micro-organisms.

3.1 Viruses

Over the period of 1975–2003 a total seven studies have been found in literature describing the removal of viruses or phages by conventional treatment (Table 3.1). Removal of both environmental pathogenic viruses and coliphages by full-scale treatment was studied by Payment *et al.* (1985; 1993), Havelaar *et al.* (1995) and Hijnen *et al.* (2003). Foliguet et Doncoeur, 1975 describe a pilot plant study after poliovirus removal. The other references (Guy *et al.*, 1977; Harrington *et al.*, 2003) present studies after coliphage removal in pilot plants. The average FS-index was 3.6, indicating a moderate level of representation of full-scale conditions (Table 3.1). Based on these studies MEC for viruses was 3.0 log with a wide range of DEC-values of 1.2–5.3 log (Figure 3.1).

Low removal (1.2–1.7 log) was calculated for the removal of:

- environmental enteroviruses and coliphages (somatic or FRNA) by a full-scale plant in the Netherlands (Havelaar *et al.*, 1995; Hijnen *et al.*, 2003) with flocculation in open basins;
- environmental coliphages by a pilot plant with sedimentation (Guy *et al.*, 1977).

Both systems used sand filtration as secondary floc-removal. High removal was observed for environmental enteric viruses in full-scale plants with little process information (Payment *et al.*, 1985;1993) and for seeded poliovirus in a pilot plant (Foliguet et Doncoeur, 1975). One of the treatment plants studied by Payment used pre-oxidation before the coagulation process which may have caused an additional inactivation effect on viruses and phages.

In the reviewed studies essential information on the three major process conditions (type of flocculation and filtration and the hydraulic retention time HRT) was frequently missing (Figure 3.1; ni = no information). The available information on these conditions did not show a relation with DEC.

Table 3.1 MEC of coagulation/floc-removal and rapid filtration (Conventional treatment)

Organisms	Data characteristics			MEC (log)		
	Studies	Data	FS-index*	Average	P50	Range
Viruses	7[a,b,d,f,j,z,A]	69	3.6	3.0 (±1.4)	2.5	1.2–5.3
Bacteria[b]	7[d,g,j,q,s,z,A]	54	3.1	2.1 (±0.8)	2.1	1.0–3.4
Bacterial spores	11[f,g,j,m,n,q,r,s,t,v,z]	62	4.7	2.4 (±0.9)	2.1	1.4–4.7
Cryptosporidium	15[e,f,h,i,l,k,o,p,t,v,w,x,y,z,A]	162	3.7	3.2 (±1.3)	2.9	1.4–5.5
Giardia	8[c,e,f,h,l,o,u,z]	67	4.3	3.4 (±0.9)	3.3	2.1–5.1

[a]Foliguet et Doncoeur, 1975; [b]Guy *et al.*, 1977; [c]Logsdon *et al.*, 1981; [d]Payment *et al.*, 1985; [e]LeChevallier *et al.*, 1991: [f]Payment *et al.*, 1993; [g]Hijnen *et al.*, 1994; [h]Nieminski *et al.*, 1994; [i]West *et al.*, 1994; [j]Havelaar *et al.*, 1995; [k]Nieminski *et al.*, 1995; [l]Patania *et al.*, 1995; [m]Rice *et al.*, 1996; [n]Hijnen *et al.*, 1997; [o]States *et al.*, 1997; [p]Yates *et al.*, 1997a; [q,r,s]Hijnen *et al.*, 1998a,c,e; [t]Nobel *etal.*, 1999; [u]Hasimoto *et al.*, 2000; [v]Dugan *et al.*, 2001; [w]Cornwell *et al.*, 2001; [x]Hasimoto *et al.*, 2001; [y]Akiba *et al.*, 2002; [z] Hijnen*et al.*, 2003; [A]Harrington *et al.*, 2003 *FS = full-scale index; the higher the number, the more equivalent with full-scale situation; [b]indicator bacteria (*E. coli,* coliforms, faecal streptococci)

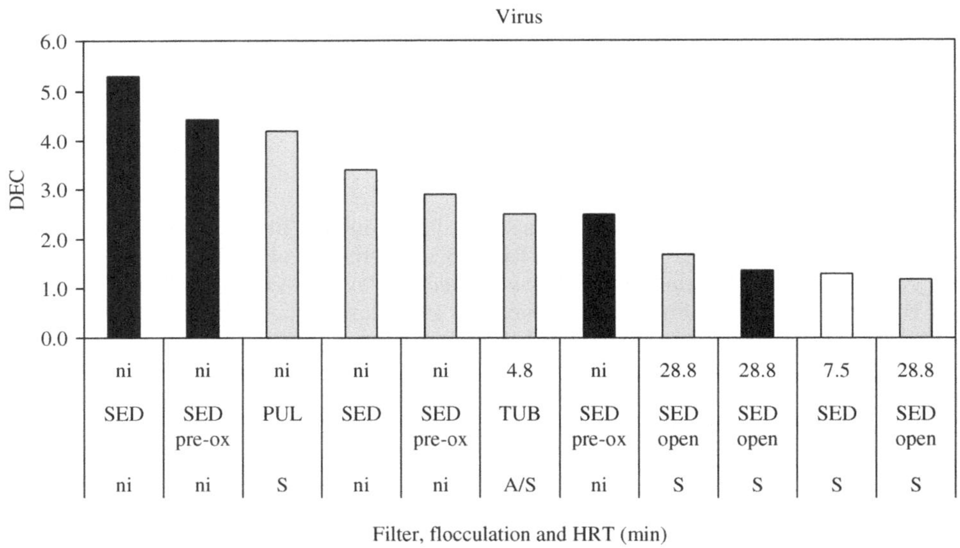

Figure 3.1 DEC of the conventional treatment for viruses arranged in declining order (Black, grey and white bars: FS-index of 5, 4/3 and 2/1, respectively); Filter, S = sand, A/S = anthracite/sand; flocculation, SED = sedimentation, PUL = pulsator, pre-ox = pre-oxidation; HRT, hydraulic retention time; ni = no information

3.2 Bacteria and bacterial spores

For the elimination of environmental thermotolerant coliforms or enterococci by full-scale conventional treatment processes six of a total of seven studies have been evaluated (Havelaar *et al.*, 1995; Hijnen *et al.*, 1994; 1998a,b; 2003; Payment *et al.*, 1985). Harrington *et al.* (2003) quantified the removal of pathogenic bacteria *E. coli* O157 and *Aeromonas hydrophyla* in a pilot plant study. The overall FS-index is 3.1. Based on these studies bacteria are removed about 1 log less by conventional treatment than viruses (Table 3.1) with a lower variation; DEC-values ranged from 1–3.4 log (Figure 3.2).

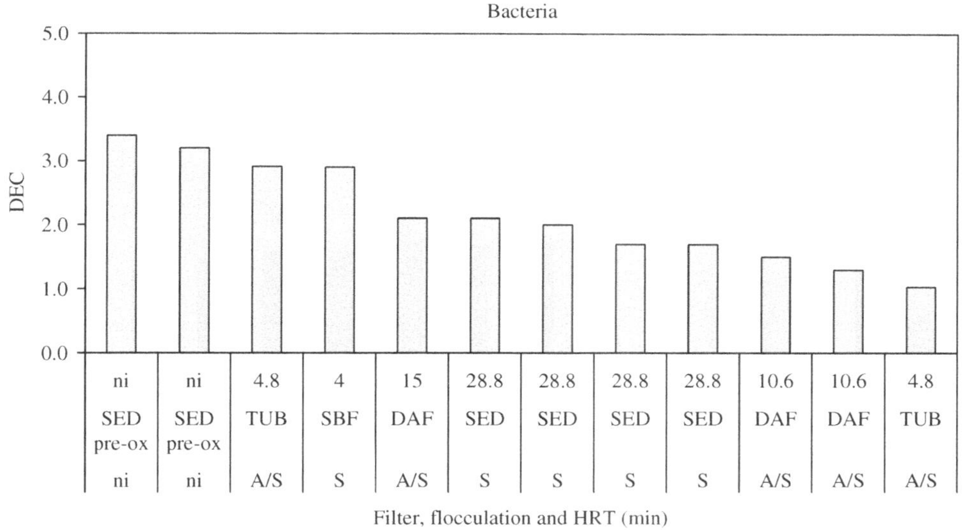

Figure 3.2 DEC of the conventional treatment for bacteria arranged in declining order (Black, grey and white bars: FS-index of 5, 4/3 and 2/1, respectively); Filter, S = sand, A/S = anthracite/sand; flocculation, SED = sedimentation, PUL = pulsator, DAF = dissolved air flotation, pre-ox = pre-oxidation; HRT, hydraulic retention time; ni = no information

High removal of around 3 log (Figure 3.2) was observed by Payment *et al.*, 1985, Harrington *et al.*, 2003 and Hijnen *et al.*, 1998a. These high values were achieved by full-scale system with pre-oxidation (Payment), a tube-settler on pilot plant-scale (Harrington) and a full-scale sludge blanket flocculator (Hijnen), respectively. Low (1 log) was the DEC of a pilot plant consisting of a tube settler and dual media filter (Anthracite/sand; A/S) for *E. coli* O157 tested by Harrington *et al.*, 2003. The latter author observed a DEC of 3.0 for *A. hydrophilic* in the same pilot plant which indicates that removal of different types of bacteria by conventional treatment may vary significantly. From the data of the full-scale studies however hardly any difference in removal of environmental coliforms and enterococci was observed (Figure 3.3). *E. coli* O157 removal by conventional treatment was not directly compared with the removal of environmental indicator bacteria like thermotolerant coliforms and enterococci. Hence, it remains the question whether these indicator bacteria can be used as surrogates to quantify removal of pathogenic *E. coli* O157.

Besides the influence of pre-oxidation on the DEC for bacteria, influence of the other process conditions (flocculation and HRT) was not observed (Figure 3.2).

The removal of bacterial spores by conventional treatment was studied in eleven separate studies (Dugan *et al.*, 2001; Havelaar *et al.*, 1995; Hijnen *et al.*, 1994, 1997, 2003, 1998a,b,c; Nobel and Medema, 1999; Payment and Franco, 1993; Rice *et al.*, 1996). Rice and Dugan studied aerobic spores and the other authors spores of sulphite-reducing clostridia (SSRC) or *C. perfringens*. All data, except the data of Dugan and part of the data of Rice, were collected for environmental spores in full-scale systems. Therefore the average FS of 4.7 (Table 3.1) is relatively high.

Based on these data the MEC is 2.4 log (arithmetic average), but based on the distribution of values (Figure 3.3) the median value (P50) of 2.1 log is a better value for describing MEC. This value is similar to that calculated for bacteria. High elimination efficiency (4.7 log) was observed for the anaerobic spores in full-scale treatments examined by Payment (1993) with little process information and by Dugan *et al.* (2001) for aerobic spores in a pilot plant (3.1 log). Hijnen *et al.* (1998a) observed the lowest DEC for anaerobic spores removed by a sludge blanket clarifier (SBC) followed by a sand filtration with low HRT of 4 minutes (1.4 log).

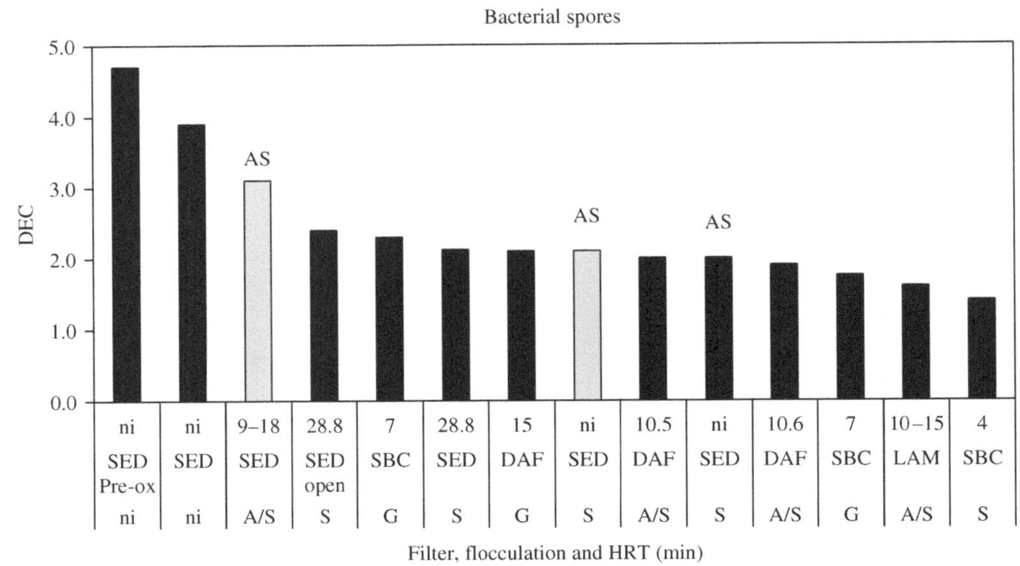

Figure 3.3 DEC of the conventional treatment for bacterial spores (aerobic (as) and clostridial spores (rest)) arranged in declining order (Black, grey and white bars: FS-index of 5, 4/3 and 2/1, respectively); Filter, S = sand, A/S = anthracite/sand; flocculation, SED = sedimentation, PUL = pulsator, DAF = dissolved air flotation, pre-ox = pre-oxidation; HRT, hydraulic retention time; ni = no information

3.3 *Cryptosporidium* and *Giardia*

In the last decade of the 20th century surveys of waterborne diseases showed that protozoan oocysts (especially *Cryptosporidium* and *Giardia*) pose a serious threat for the drinking water industry in the developed countries. Because of the high resistance of these oocysts to disinfection, physical processes like coagulation/flocculation and filtration (conventional treatment) are the most significant barriers in water treatment.

A total of 15 studies collected data for *Cryptosporidium* oocyst and 8 studies for *Giardia* cyst removal (Table 3.1). Removal of indigenous organisms under full-scale conditions was studied by Hashimoto *et al.* (2001) and Hijnen *et al.* (2003), LeChevallier *et al.* (1992), Payment and Franco, 1993, States *et al.* (1997), Yates *et al.* (1997), . The other studies were conducted mostly under pilot plant conditions with spiked oocysts (Logsdon *et al.* (1981); West *et al.* (1994); Patania *et al.* (1995); Nieminski *et al.* (1994; 1995); Cornwell *et al.* (2001); Dugan *et al.* (2001; 2004); Harrington *et al.* (2003)). Microbiota (including algae) counting microscopically (MPA) as a tool to assess plant performances was studied by Hancock *et al.* (1996) in a survey and revealed an average DEC of 2.9 log (n = 110) for 55 full-scale conventional water treatment plants in the US. The range of DEC was large (0–6.2 log). Because MPA covers small as well as larger microbiota and the study did not describe the process configuration of the examined plants, these data were not included in the MEC calculations for both protozoa. The algae data of Nobel and Medema (1999) and Akiba *et al.* (2002) were included because they studied the removal of small algae by conventional treatment as surrogates for protozoan oocysts.

The FS-index for *Cryptosporidium* and *Giardia* was 3.7 and 4.3, respectively, indicating a higher agreement with full-scale conditions compared to viruses and bacteria (Table 3.1).

Removal of micro-organisms in conventional treatment will largely be ruled by physical/chemical processes like surface interactions between particles (other colloids or filter materials), sedimentation and physical straining. In these processes higher removal may be expected for the larger sized organisms and this was confirmed by the higher MEC for both protozoan oocysts compared to the bacteria, spores and viruses (Table 3.1). MEC for *Cryptosporidium* and *Giardia* was 3.2 and 3.4, respectively. The range of DEC-values was in the same order of magnitude as observed for viruses.

A clear indication about the influence of the process configuration (type of flocculation and filtration) on MEC was not observed for both organisms (Figure 3.4 and 3.5). High was the removal assessed by Payment (1993) in a full-scale plant with Pre-oxidation and relative low was the *Giardia* removal observed by LeChevallier in full-scale treatment plants.

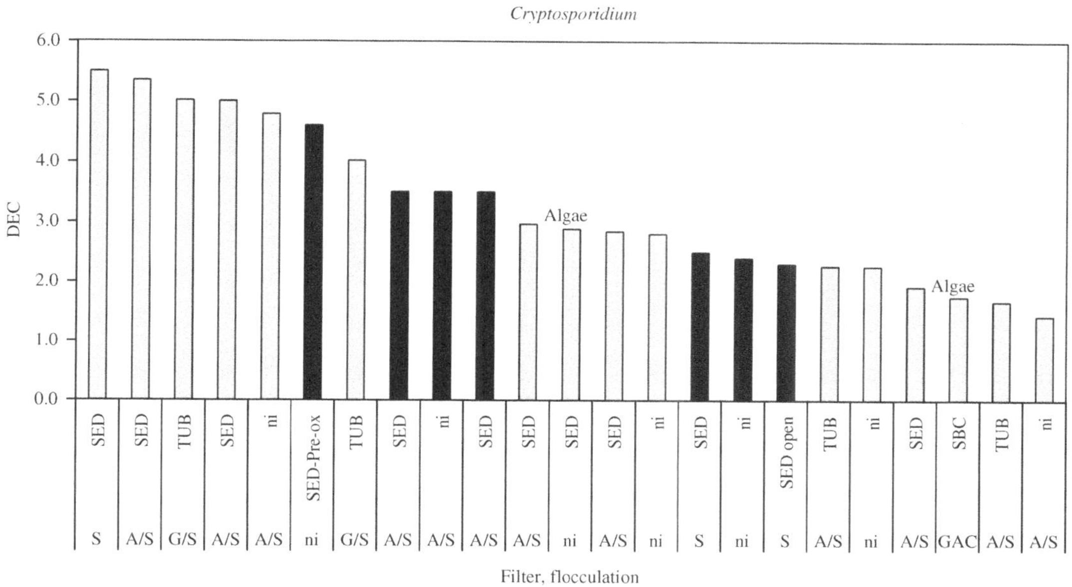

Figure 3.4 DEC of the conventional treatment for Cryptosporidium oocysts arranged in declining order (Black, grey and white bars: FS-index of 5, 4/3 and 2/1, respectively); Filter, S = sand, A/S = anthracite/sand, GAC = granular activated carbon filtration; flocculation, SED = sedimentation, TUB = tube settlers, DAF = dissolved air flotation, pre-ox = pre-oxidation; ni = no information

The algae studies were included in the *Cryptosporidium* data and indicated separately in Figure 3.4. These data showed that removal of these organisms by full-scale plants was in the same order of magnitude as observed for both protozoan oocysts.

3.4 Discussion

In some of the evaluated studies described in this chapter effects of process conditions on the overall removal efficiency of conventional treatment was presented. Furthermore, several authors studied more than one micro-organism with the objective to investigate potential indicators or surrogates for pathogen removal by conventional treatment. This paragraph describes the major findings of these studies. Detailed effects of process conditions of the two separate processes, coagulation/flocculation and filtration, will be described in the chapters specifically describing removal efficiency of these separate processes.

In the study methodology described in Chapter 2 a weighing of DEC-values was introduced based on similarity with full-scale conditions. The significance of this method as well as the variation in observed DEC-values presented in literature will be discussed subsequently.

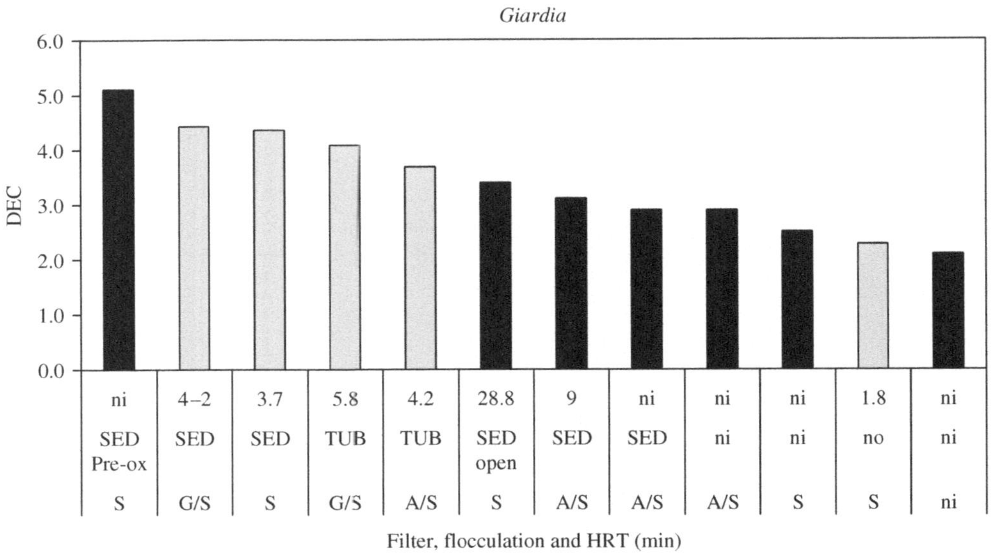

Figure 3.5 DEC of the conventional treatment for Giardia cysts arranged in declining order (Black, grey and white bars: FS-index of 5, 4/3 and 2/1, respectively); Filter, S = sand, A/S = anthracite/sand, GAC = granular activated carbon filtration; flocculation, SED = sedimentation, TUB = tube settlers, DAF = dissolved air flotation, pre-ox = pre-oxidation; ni = no information

3.4.1 Process conditions

No effect of process design characteristics like type of flocculation (sedimentation, lamellae separation, flotation) or filtration (dual media, sand filtration) on the overall removal of micro-organisms by conventional treatment has been observed from this evaluation (Figure 3.1–3.5). But some individual studies clearly demonstrate effects of operational conditions on the removal.

3.4.1.1 Effect of coagulation conditions

Influence of pre-oxidation: First of all, the results derived from the accumulated studies described in the former paragraphs, indicated that pre-oxidation is favourable for removal efficiency of conventional treatment. Additional inactivation of susceptible organisms is one of the explanations for this effect, but also an enhancement of coagulation may contribute to this enhancement of the removal (Jekel, 1994; Saunier *et al.*, 1983;).

Influence of chemical dose: The significance of the optimal dose of chemicals was demonstrated by Dugan *et al.* (2001) in a pilot plant study. In a period with under dosing of the coagulants *Cryptosporidium*, aerobic spores and particles were clearly removed less efficient than in the period with optimal coagulant dose (Figure 3.6).

Influence of pH: Electrostatic forces play an important role in the attachment and removal of microbial colloids in conventional treatment. These forces are being influenced by the iso-electrical point (IEP) of the colloids and the surface of the granular material (collector). IEP is the pH where the surface charge is zero (un-charged). IEP of most micro-organisms and filter grains is low; therefore at pH in water treatment (neutral) both surfaces will be charged negatively and endure high repulsive forces (non attachment). At low pH, charge diminishes which is favourable for attachment. Based on this consideration more removal is to be expected at low pH. For poliovirus removal, C. parvum, E. intestinalis, A. hydrophyla, E. coli O157 and MS2 phage by coagulation and filtration both Foliguet and Doncoeur (1975) and Harrington *et al.* (2003) showed no clear enhancement of removal at low pH in the range of 5–8 (Figure 3.7 and 3.8).

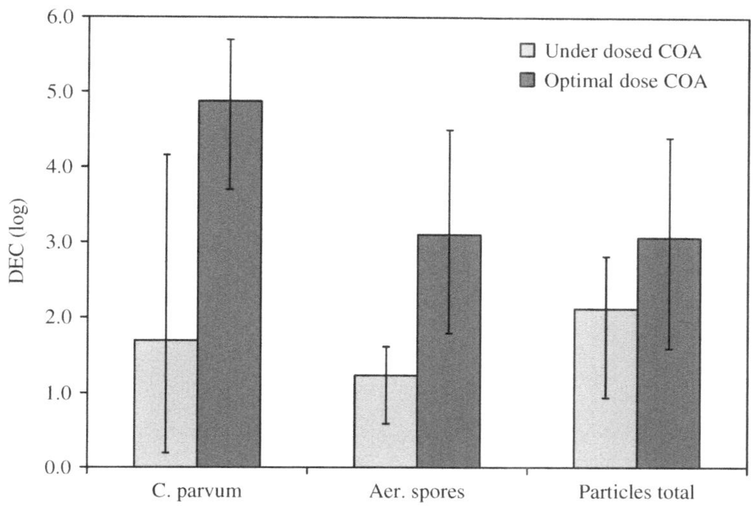

Figure 3.6 The DEC of Cryptosporidium, aerobic spores and particles during a period with low and optimal dosing of coagulants (vertical bars: range of values; from Dugan *et al.*, 2001)

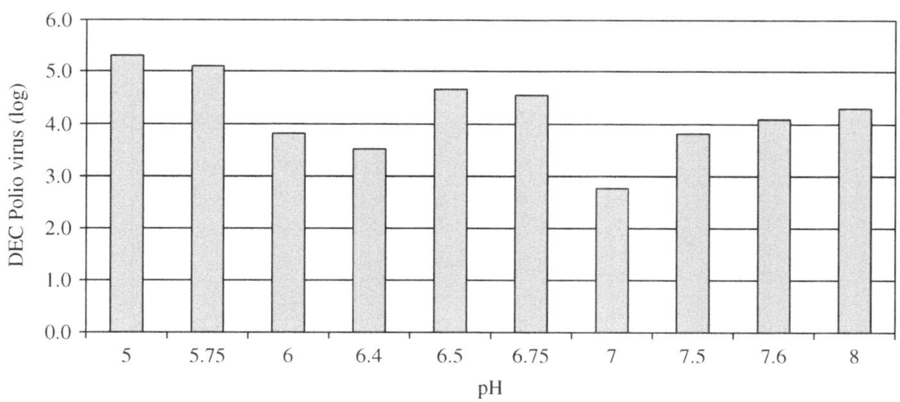

Figure 3.7 DEC of coagulation (flocculation by pulsator) and sand filtration for polioviruses as function of the pH during coagulation (Foliguet and Doncoeur (1975))

In Chapter 4 the effect of coagulation conditions on the removal of micro-organisms by this separate process will be discussed more in detail.

3.4.1.2 Effect of filtration conditions

As described in the former paragraphs, no clear effect of filter design (materials, HRT) on the overall removal by conventional treatment was observed. Filtration in the conventional treatment is an intermittent process, since the filter is back washed periodically to reduce head loss and remove accumulated suspended solids. It is well known that colloid removal at the start and the end of filter cycles is less efficient due to reduction in filtration coefficient and breakthrough behaviour, respectively. But this is not observed in every study.

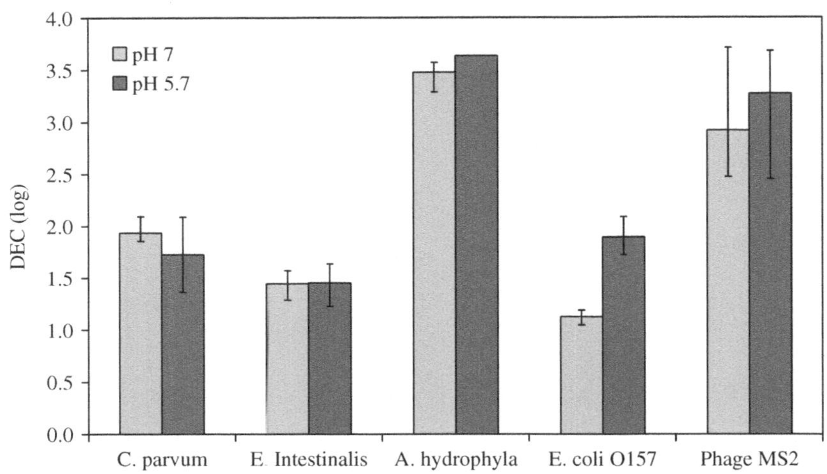

Figure 3.8 DEC of coagulation (flocculation by sedimentation) and filtration (sand, dual media, tri-media) for protozoan oocysts and bacteria (Harrington *et al.*, 2003)

Harrington *et al.* (2003) for instance demonstrated high removal after back washing and a reduction in removal efficiency for protozoa, bacteria and phages during the period of a filter cycle (Figure 3.9).

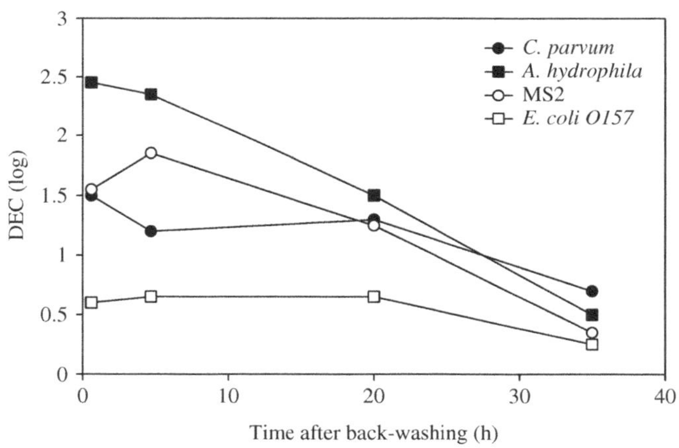

Figure 3.9 The removal of several micro-organisms by coagulation/filtration during the period of a filter cycle (9.7 m/h and HRT of 4.8 minutes; from Harrington *et al.*, 2003)

Patania *et al.* (1995) showed reduced DEC of conventional treatment for protozoan oocysts during the maturation period of the filter compared to the DEC assessed during stable operation (turbidity <0.1; Figure 3.10). But these data also show that DEC for *Cryptosporidium* assessed during maturation were in 35% of the observations higher than DEC assessed in the subsequent stable period of the run (Figure 3.10). The effect of the filter cycle on removal by filtration will be discussed in more detail in the chapter evaluating the removal efficiency of filtration processes.

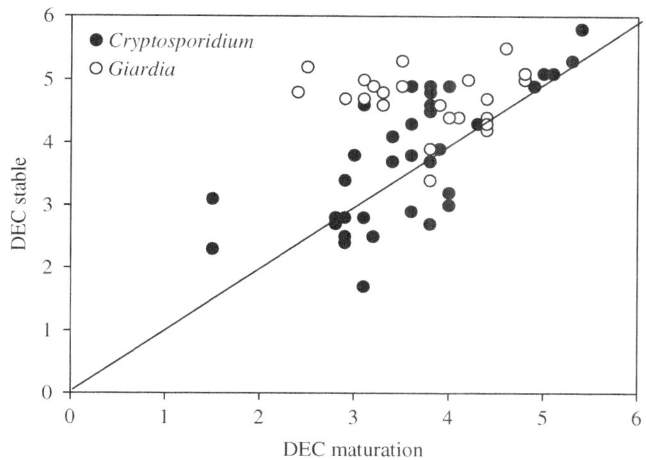

Figure 3.10 The removal of protozoan oocysts by coagulation/filtration by three different pilot plants during the period of maturation and the period of stable operation (turbidity <0.1; Patania *et al.*, 1995)

3.4.2 FS-index and variation

To investigate the significance of the FS-index on the calculated MEC-values, the FS-index for removal of indigenous viruses and protozoa by full-scale processes (FS = 5) was doubled (Table 3.2). The MEC is hardly influenced by this variation in FS-value. This was because DEC-values found in these full-scale studies are equally distributed over the total data-sets (Figure 3.1, 3.4 and 3.5).

Table 3.2 Influence of FS-index on MEC of coagulation/floc-removal and rapid filtration (Conventional treatment)

	FS-index full-scale study 5 (Table 3.1)	**MEC FS = 5**	FS-index full-scale study 10	**MEC FS = 10**
Viruses	3.6	**3.0**	5.5	**3.1**
Cryptosporidium	3.7	**3.2**	5.3	**3.2**
Giardia	4.3	**3.4**	7.3	**3.3**

In Figure 3.11 the frequency distribution of the DEC-values for the different micro-organisms is presented. This figure showed that the protozoan (oo)cysts were removed more efficiently than the other micro-organisms. For viruses most DEC-values were found in the ranges 1–1.5 log and 2.5–3 log, but the overall range of DEC-values was large. The DEC-values for bacteria was more equally distributed between 1–3.5 log, while for the spores a clear peak was observed in the interval 2–2.5 log.

Though DEC for *Cryptosporidium* showed a wide range of values, a small peak was observed in the interval 2.5–3 log. For *Giardia* DEC-values showed two peaks of DEC and a higher minimal DEC-value than observed for *Cryptosporidium*.

3.4.3 Surrogates for conventional treatment

To assess the elimination capacity of treatment processes for pathogenic viruses, bacteria and protozoan oocysts the use of surrogates is required. In several studies the removal of several micro-organisms by conventional treatment is determined simultaneously (Table 3.3). For some surrogates application can be discussed based on the average DEC simultaneously assessed and for some surrogates data are collected to study the similarity of breakthrough behaviour.

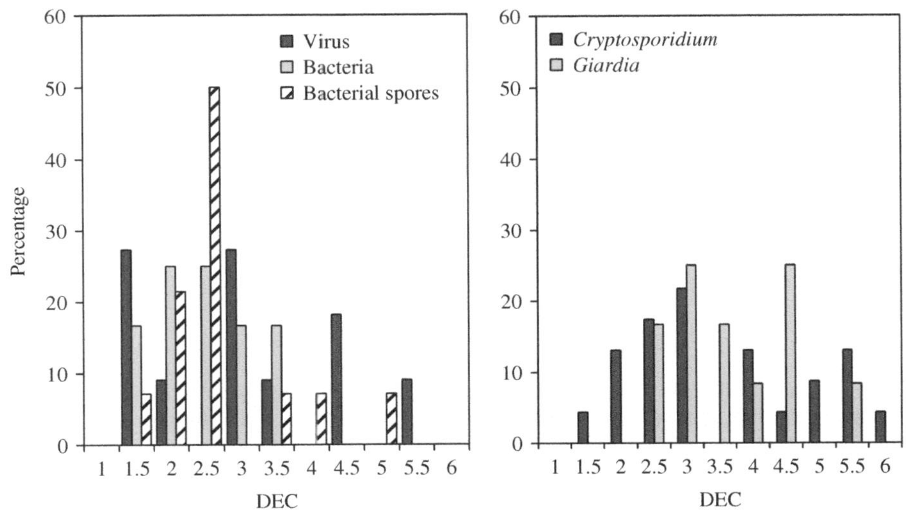

Figure 3.11 Frequency distribution of the DEC-values observed in literature (x-axis is maximum value of each interval)

Bacteria. No studies have been found comparing removal of indicator bacteria with pathogenic bacteria by conventional treatment. A number of studies indicated that coliforms were removed to the same extent as enterococci (Table 3.3). These DEC-values were significantly higher compared to the DEC described by Harrington *et al.* (2003) for *E. coli* O157. Because thermotolerant coliforms (Coli44) and *E. coli* O157 belong to the same species it is not likely that DEC assessed under similar conditions show the same difference as presented here. Future experiments will have to elucidate the use of Coli44 as surrogate for this enteropathogenic bacterium.

Bacteriophages. Three studies compared the elimination of bacteriophages and human enteric viruses by conventional treatment simultaneously (Table 3.3) and the results were inconclusive. Two studies showed somewhat more removal of bacteriophages than the enteric viruses and another study showed the opposite. Thus, more research is needed to show the applicability of phages as surrogates for virus removal.

Cryptosporidium and Giardia. For these protozoan oocysts a number of potential surrogates have been mentioned and evaluated in literature. Turbidity and particle counting as on-line parameters (e.g. Dugan *et al.*, 2001; LeChevallier *et al.*, 1991; Nieminski *et al.*, 1994;1995), bacterial spores (Dugan *et al.*, 2001; Hijnen *et al.*, 1997; Payment and Franco, 1993; Rice *et al.*, 1996) and algae (Akiba *et al.*, 2002; Nobel and Medema, 1999).

In the six studies comparing turbidity reduction with protozoan oocyst removal all studies but one presented lower DEC for turbidity than for the *Cryptosporidium* (Table 3.3). In the study of Dugan *et al.* (2001) a statistical significant correlation ($P<0.05$) between the removal of *Cryptosporidium* and turbidity (0.42) and total particle count (0.51). In two studies regression analysis of the data indicated also lower DEC for turbidity than for *Cryptosporidium* (Figure 3.12).

Turbidity, just as particle count, are sum-parameters influenced by a mix of all kinds of suspended solids which vary as a result of preceding processes. Therefore these observations are not surprising. Based on the DEC-values presented in Table 3.3 and Figure 3.13, particle counting showed to be a better surrogate for protozoan removal than turbidity, but still oocysts removal is generally higher than particle removal.

Table 3.3 Elimination of micro-organisms (pathogenic, indicator, surrogates) and physical parameters (surrogates) by conventional treatment

Author	Bacteria			virus		Protozoa					
	Coli44	Enterococ.	E.coli O157	Phages	HEV	Spores	Algae	Particles	Turbidity	Crypto	Giardia
Payment 1985	3.2	3.4			4.4						
Hijnen 1994	1.5	1.3				1.2					
Hijnen 1998c	2	2.1				2.1					
Havelaar 1995	2.0	2.1		1.7	1.4	2.1					
Hijnen 2003	1.7	1.7		1.2		2.4				2.3	
Payment 1993				3.4	5.3	3.8					
Payment 1993 pre-ox				2.9	2.5	4.7				4.6	5.1
Harrington 2003			1	2.5						1.7	
Dugan 2001 (low dose)						3.1		3.1	2.2	5	
Dugan 2001 (opt. dose)						1.2		2.1	1.4	1.7	
Patania 1995 (Azusa)								3	2	>5.5	
Patania 1995 (Con.Costa)								2	1.5	4	
Patania 1995 (Tualatin)								1.5	3	5	
Nieminski 1995								2.7–2.8		2.6	3.2
Nobel 1999						1.8	1.8				
Akiba 2002							2.9				
Hashimoto 2001									3.1	2.5	2.5
LeChevallier 1992								0–4		0–2.5	0–4.0
									0–3.0	0–4.0	0–4.0

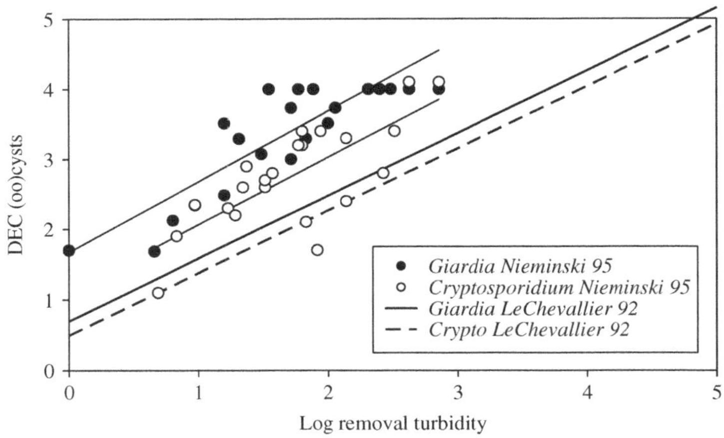

Figure 3.12 Regression analysis of DEC of conventional treatment for protozoan oocysts and turbidity (from LeChevallier *et al.*, 1992; Nieminski *et al.*, 1995)

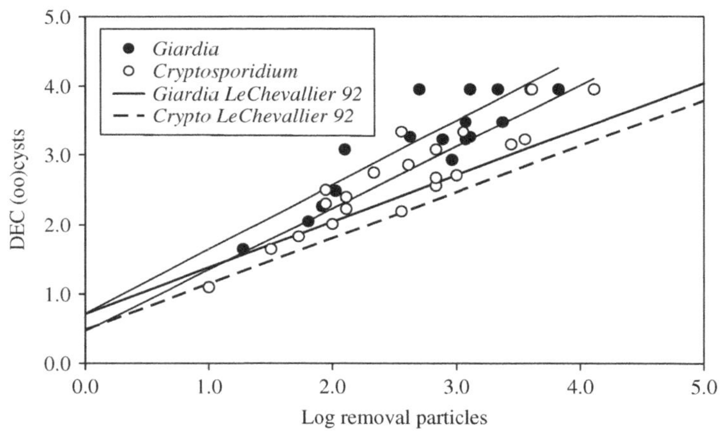

Figure 3.13 Regression analysis for the removal of protozoan oocysts and particles by conventional treatment simultaneously assessed by LeChevallier *et al.* (1992; indigenous protozoa and particles >5 um) and Nieminski *et al.* (1995; spiked (oo)cysts and particles 7–11 and 4–7 um for Cryptosporidium and Giardia)

From the overall evaluation presented in Table 3.1 it can be concluded that MEC of conventional treatment for bacterial spores is 0.8–1.0 log lower than MEC of this process for protozoan oocysts. In three of the four studies where these organisms have been studied simultaneously, however, the DEC-values were more in the same order of magnitude (Table 3.3). Dugan 2001 showed a correlation between Crypto and bacterial spores ($r^2 = 0.67$).The removal of algae by conventional treatment presented by Akiba 2001 and Nobel 1999 was in the same order of magnitude as the removal of the other surrogates.

3.5 Conclusions and further research

Although probably not all published literature is included in this evaluation, it covers all references used by Haas *et al.* (2001) and LeChevallier (2001). The evaluation of conventional treatment processes showed that this process is an important barrier in treatment to reduce the concentration of micro-organisms (viruses, bacteria and bacterial spores and (oo)cysts of *Cryptosporidium* and *Giardia*).

- The calculated MEC for viruses and protozoan (oo)cysts are in the same order of magnitude and range between 3.0 and 3.4 log. Bacteria and bacterial spores are removed 1 log less.
- The FS-index of the evaluated data on bacterial removal was high and moderate for respectively viruses and protozoan (oo)cysts. Because the DEC-values of processes with a high FS-index were distributed evenly over the data bases the impact of the weighing on the MEC-values is negligible.
- Surrogates for pathogenic virus and bacteria removal showed inconclusive results for bacteriophages and thermotolerant coliforms. Correlation between protozoan oocyst removal and removal of turbidity, particles and spores have been found and indicate that these surrogates are safe surrogates: DEC of these surrogates is usually lower than DEC of protozoa.

The following effects of process conditions on DEC have been observed in literature:

- pre-oxidation enhance removal;
- under dosage of coagulant will reduce DEC;
- although pH will affect coagulation conditions, tests indicate that this effect is limited on the overall removal by conventional treatment;
- a pronounced effect of the back washing can be seen but this effect is variable and depends on the conditions in the filter (presumably the head loss and preferential flow patterns).

These effects will be discussed for the two separate processes (coagulation/filtration) in the following Chapters.

4 Coagulation and flocculation

Coagulation and flocculation is usually applied in combination with rapid sand filtration, as separate process or as in-line process in the filter bed (direct filtration). The combination of coagulation and filtration is indicated as conventional treatment and is discussed in the former chapter. A number of different flocculation systems are used in practice such as lamellae separation (LAM), dissolved air flotation (DAF), sludge blanket clarifier (SBC), tube-settlers (TUB) and sedimentation in open reservoirs or in a small scale jar-test (SED). Removal data of these processes are described per group of micro-organisms in the following paragraphs.

4.1 Viruses

A MEC for viruses of 1.8 log was calculated from a total of 89 DEC-values, calculated from eight studies (Table 4.1). In four of these studies the removal of environmental viruses and bacteriophages by full-scale processes was determined (FS-index of 5 and 3, respectively; Hijnen *et al.*, 1994 and Havelaar *et al.*, 1995; Payment *et al.*, 1985, 1993; Stetler *et al.*, 1992). Seven of the DEC-values were measured in bench lab-scale studies (FS-index of 1 and 2; Bell *et al.*, 2000; Nasser *et al.*, 1995; Rao *et al.*, 1988). The average FS-index of 3.5 is in the same order of magnitude as calculated for the other organisms and indicates a moderate level of agreement with full-scale conditions.

The variation in DEC-values reported in the different studies was high (0.2 up to >4 log; Figure 4.1). Pre-oxidation increased the removal efficiency due to the additional effect of the disinfectant when available as a free chemical or due to the increased efficacy of the coagulation process itself (among others Jekel, 1994; Saunier *et al.*, 1983). Elimination by coagulation and sedimentation in an open reservoir (lagoon coagulation) was limited, most likely due to low efficiency of sedimentation. Jar-testing is usually applied to obtain the optimal chemical dosing conditions for coagulation/flocculation processes.

The lab-scale study of Rao *et al.* (1988) determined the removal of Hepatitis A, polio and rota-virus by sedimentation (jar-test and small pilot plant installation). DEC varied between 1.0 and 1.7 log. Payment *et al.* (1985) presented a study where the removal of environmental viruses was monitored at seven full-scale water treatment plants with sedimentation as the floc-removal process. A significant monitoring program was performed (n = 25). The process conditions were not well documented and at all locations coagulation was operated with a pre-chlorination (Pre-ox). The average DEC was 2 log and ranged from 1.3 up to 2.6 log. Stetler *et al.* (1992) presented a study where the removal of environmental enteric viruses and coliphages was determined in a treatment train with coagulation (4 samples; few quantitative data). The calculated DEC of coagulation for both organisms was low (0.2, 0–0.5 log; n = 7).

In 1993 Payment published a study where the removal of several micro-organisms of faecal origin by three full-scale water treatment systems was monitored (Payment *et al.*, 1993). Coagulation and sedimentation removed human enteric viruses with 2.3 log (1.3–3.2) whereas coliphage concentrations decreased with 2.5 log (1 up to 5.3 log). Pre-ozonation increased DEC with 1.5 log for all these organisms. In the winter period of 1991 Havelaar *et al.* (1995) studied the virus removal by full-scale coagulation processes in the Netherlands. Enteroviruses were removed with 0.6 log by coagulation/sedimentation and with 1.9 log by coagulation/lamellae separation. Environmental coliphages in the water were removed by 1.9 log in a full-scale dissolved air flotation (DAF) process with a marginal chlorine dose introduced by the saturation water of the process (0.02–0.04 mg/l; Hijnen *et al.*, 1994).

Table 4.1 MEC of coagulation/floc-removal

Organisms	Data characteristics			MEC (log)		
	Studies	Data	FS-index*	Average	P50	Range
Viruses	8[b,c,e,f,g,i,j,w]	89	3.5	1.8 (±0.7)	1.7	0.2–4.3
Bacteria[b]	9[b,g,e,i,m,p,t,v,w]	101	3.0	1.5 (±0.9)	1.4	0.6–3.7
Bacterial spores	10[e,f,i,o,p,q,t,u,v,w,y]	92	4.6	1.4 (±0.6)	1.2	0.8–3.2
Cryptosporidium	14[d,f,h,k,l,n,r,s,w,x,y,z,A,B]	92	3.3	1.9 (±0.9)	1.8	0.4–3.8
Giardia	9[a,d,f,h,k,r,w,z,A]	53	3.9	1.6 (±0.9)	1.3	0.3–2.9

[a]Logsdon *et al.*, 1985; [b]Payment *et al.*, 1985; [c]Rao *et al.*, 1988; [d]LeChevallier *et al.*, 1991; [e]Stetler *et al.*, 1992; [f]Payment *et al.*, 1993; [g]Hijnen *et al.*, 1994; [h]Kelley *et al.*, 1994; [i]Havelaar *et al.*, 1995; [j]Nasser *et al.*, 1995; [k]Patania *et al.*, 1995; [l]Plummer *et al.*, 1995; [m]van Puffelen *et al.*, 1995; [n]Kelley *et al.*, 1996; [o]Rice *et al.*, 1996; [p]Gray *et al.*, 1997; [q]Hijnen *et al.*, 1997; [r]States *et al.*, 1997; [s]Edzwald *et al.*, 1998; [t,u,v]Hijnen *et al.*, 1998c,e,f; [w]Bell *et al.*, 2000; [x]Cornwell *et al.*, 2001; [y]Dugan *et al.*, 2001; [z]Edzwald *et al.*, 2001; [A]Akiba *et al.*, 2002; [B]Edzwald *et al.*, 2003; *FS = full-scale index; the higher the number, the more equivalent with full-scale situation; [b]indicator bacteria (*E. coli,* coliforms, faecal streptococci)

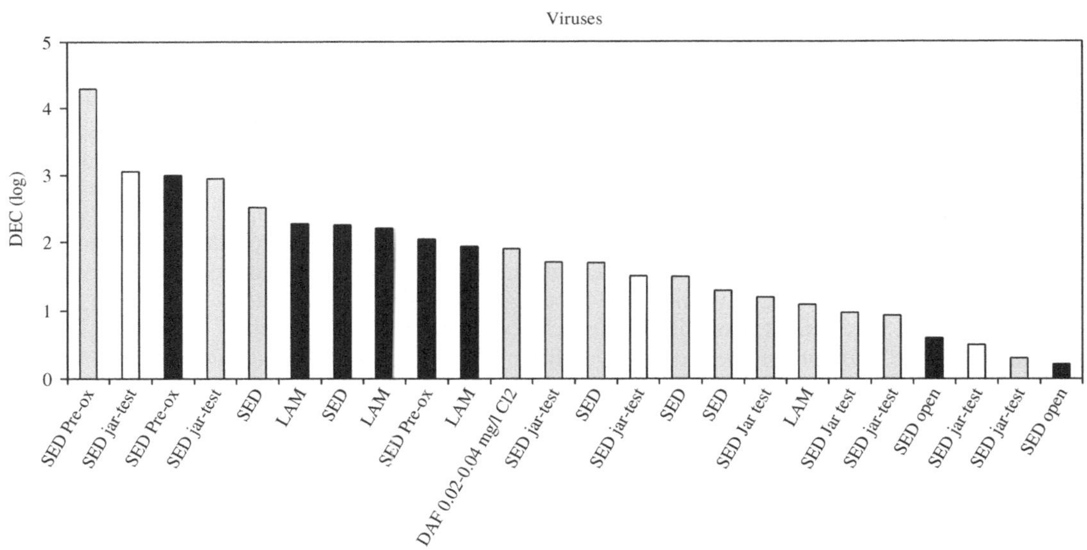

Figure 4.1 DEC of the coagulation and flocculation for viruses arranged in declining order (*Black, grey and white bars: FS-index of 5, 4/3 and 2/1, respectively*)

4.2 Bacteria and bacterial spores

Evaluation of twelve studies gave DEC-values for the elimination of indicator bacteria and bacterial spores (spores of sulphite-reducing clostridia SSRC or *Clostridium perfringens* and aerobic spores) by coagulation processes. The studies of Gray *et al.* (1999), Payment *et al.* (1985; 1993), Stettler *et al.* (1992) in full-scale plants and of Bell *et al.* (2000) and Dugan *et al.* (2001), Gray *et al.* (1999), Rice *et al.* (1996) in pilot plants and jar-tests were included and supplemented with Dutch studies. Havelaar *et al.* (1995), Hijnen *et al.* (1997; 2002; 1998b,c,d) and Van Puffelen *et al.* (1995) described the removal of thermotolerant coliforms (Coli44), enterococci, *Aeromonas* bacteria and SSRC by full-scale coagulation processes using large volume sampling. There were no studies found on the removal of pathogenic bacteria and therefore, despite the availability of full-scale data of indigenous micro-organisms, FS-index was not high (Table 4.1). Since bacterial spores are treated as separate micro-organisms within the organisms evaluated in this study and currently most data on coagulation effects came from full-scale systems monitoring indigenous spores, FS-index for these organisms was relatively high (4.3).

The MEC of 1.5 log for bacteria was calculated from results with a wide range of DEC-values (0.5 up to 3.9 log; $n = 9$; Table 4.1 and Figure 4.2).

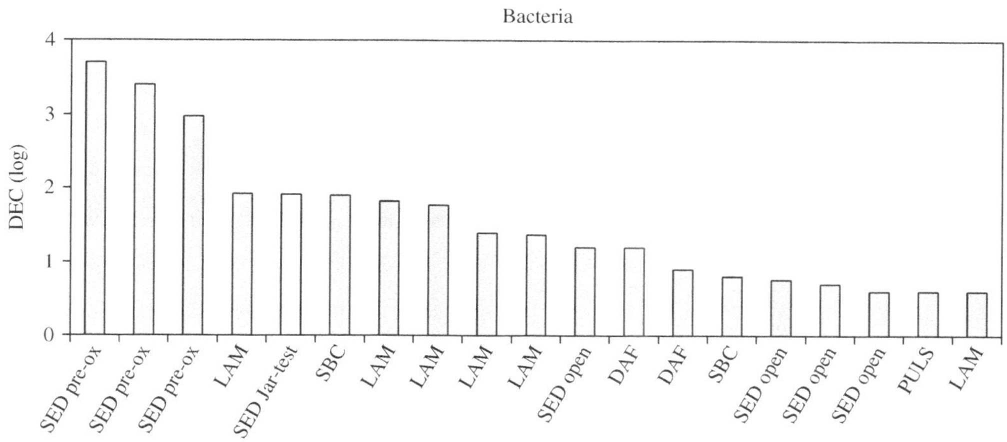

Figure 4.2 DEC of coagulation and flocculation for indicator bacteria arranged in declining order (*coliforms, E. coli, enteroccocci* and *Aeromonas (DAF)* (*Black and white bars: FS-index of 5 and 2/1 respectively*)

Bacterial spores of *Clostridium* species (*C. perfringens* or SSRC) or *Bacillus* species (aerobic spores) were eliminated by coagulation/flocculation with a MEC of 1.4 (Table 4.1) and the distribution of the DEC-values was in the same range as observed for bacteria (compare Figures 4.2 and 4.3).

High DEC-values (>3 log) were observed by Payment *et al.* (1985; 1993) for full-scale coagulation with pre-oxidation and also jar-tests showed high efficiency of removal for these micro-organisms. The latter observation indicates that chemical dosing conditions is of importance for the elimination of these micro-organisms. Relatively low was the DEC (<1 log) for full-scale sedimentation processes (Figures 4.2 and 4.3). The DEC of Dissolved Air Flotation (DAF) and Sludge Blanket Clarifier processes (few data) was close to the calculated MEC-values for bacteria and spores, indicating a more efficient elimination in this kind of flocculation processes.

4.3 *Cryptosporidium* and *Giardia*

Fifteen studies were evaluated for protozoan (oo)cyst removal (*Cryptosporidium* and *Giardia*; Table 4.1) with a total of 92 and 53 data points, respectively. Kelley (1995; 1996), LeChevallier *et al.* (1991), Payment and Franco (1993) and States (1997) determined the elimination of environmental (oo)cysts by full-scale coagulation processes. Logsdon *et al.* (1985) studied the elimination of *Giardia* and Cornwell *et al.* (2001) and Dugan *et al.* (2001), Edzwald and Kelley (1996), Edzwald *et al.* (2001;2003), Patania *et al.* (1995) of *Cryptosporidium* spiked to the influent of a pilot plant. Plummer *et al.* (1995), Edzwald and Keley (1998) and Bell (2001) studied the process on bench-scale level (jar-tests) to determine the removal of spiked *Cryptosporidium* oocysts. FS-index of these data sets (3.3 and 3.9) was in same order of magnitude as for the virus and bacteria studies. Most studies have been conducted with indigenous or lab-cultured (oo)cysts and only few studies have been used where algae removal was measured as surrogates (Akiba *et al.*, 2002).

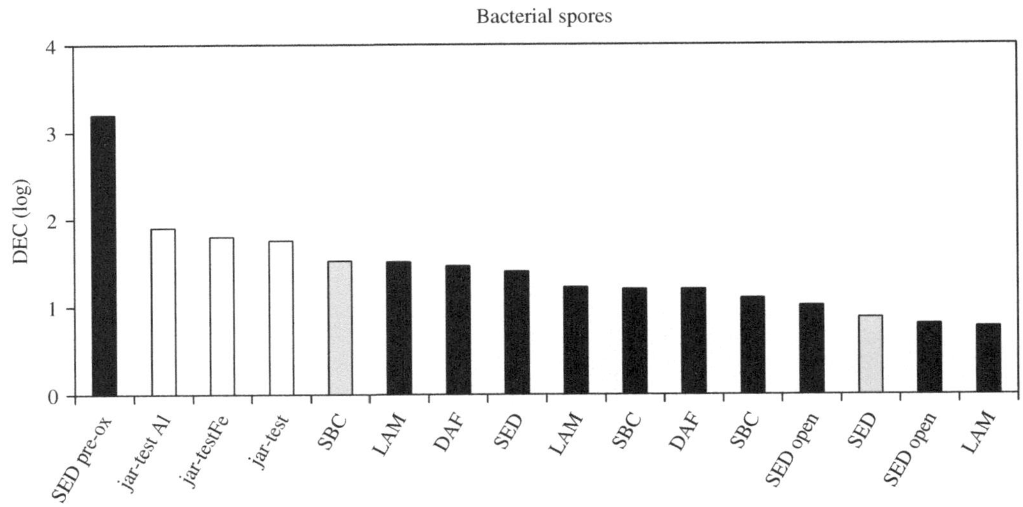

Figure 4.3 DEC of coagulation and flocculation for bacterial spores (spores of sulphite-reducing clostridia SSRC) *arranged in declining order (Black and white bars: FS-index of 5 and 2/1 respectively)*

These data yielded a MEC of 1.9 and 1.6 log for *Cryptosporidium* and *Giardia*, respectively (Table 4.1) with a similar range of DEC-values as observed for the other organisms. High DEC-values for both organisms were observed for processes with pre-oxidation and assessed with spiking experiments (FS < 5; grey bars in Figures 4.4 and 4.5). Most of the full-scale studies (FS = 5; black bars) showed lower DEC-values. Haas *et al.* (2001) calculated a MEC of 1.3 (SD = 0.99) and 2.3 (SD = 0.8) log for *Cryptosporidium* removal by coagulation followed by sedimentation and DAF, respectively. These values were calculated from a literature review (only A quality data as defined in this study; different qualification than used in this study) and were in the same range as presented in Table 4.1 and Figure 4.4. This is plausible because part of the reviewed literature was similar (more data in the present study).

4.4 Discussion

4.4.1 Process conditions

The efficacy of coagulation and flocculation processes for the removal of micro-organisms well influenced by several conditions. The nature and concentrations of the used chemicals, pH, temperature, concentration of suspended solids in the treated water, mixing mode, type and volume load of the following flocculator. These aspects have been studied in literature related to the physical water quality parameters like turbidity. Studies focused on the effects of these conditions on the removal of micro-organisms are limited. A number of these studies will be presented in the following text.

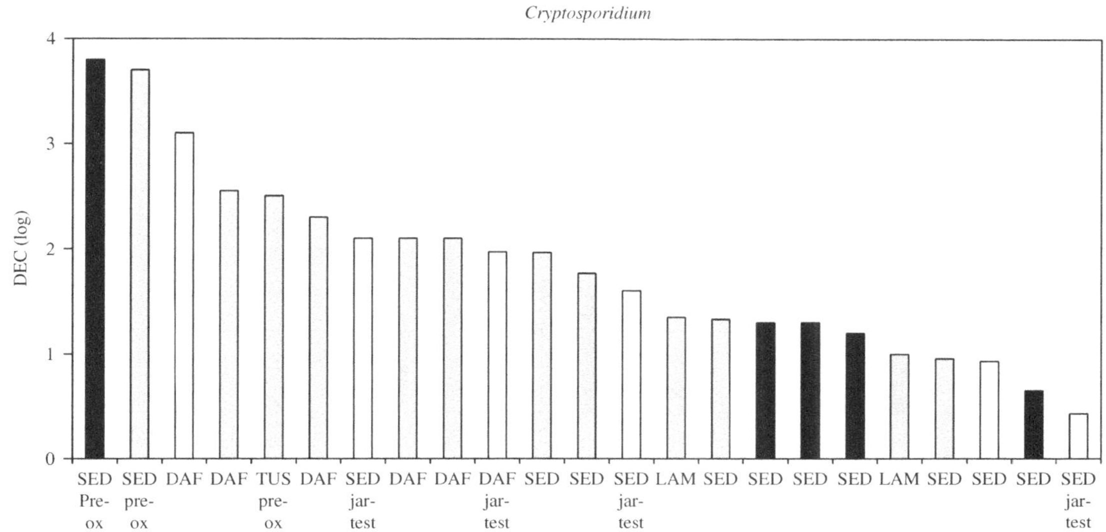

Figure 4.4 DEC of coagulation and flocculation for oocysts of *Cryptosporidium* arranged in declining order (*Black, grey and white bars: FS-index of 5, 4/3 and 2/1 respectively*)

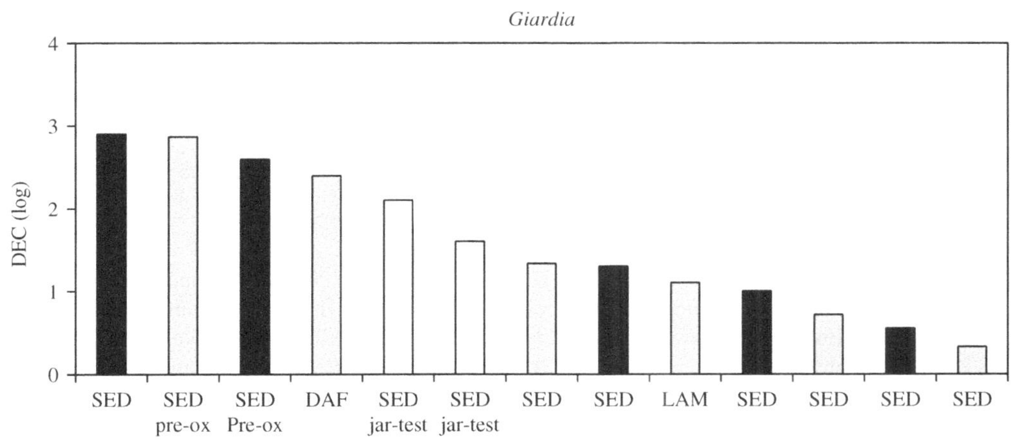

Figure 4.5 DEC of coagulation and flocculation for cysts of *Giardia* arranged in declining order (*Black, grey and white bars: FS-index of 5, 4/3 and 2/1 respectively*)

4.4.1.1 Chemicals

The most commonly used coagulants are Aluminium (Al) and Iron (Fe) hydroxides. The optimal dosage depends on the water chemistry and in Jar-tests the dosage conditions for the specific location is determined based on the lowest residual metal concentration and turbidity. For virus removal Rao *et al.* (1988) demonstrated that to a certain extent there was a relationship between the coagulant dosage and the removal efficiency of viruses and turbidity (Figure 4.6).

Nasser *et al.* (1995) showed the enhanced effect of the use of an coagulant aid (poly-electrolyte; cationic) on the removal of MS2 and Hepatitis A-virus (Table 4.2).

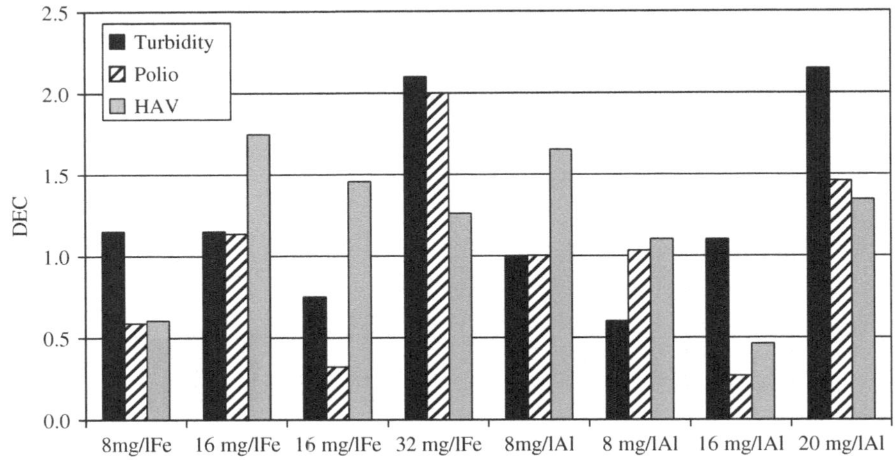

Figure 4.6 DEC values for the removal of turbidity, poliovirus and Hepatitis A virus in a jar-test with different coagulant dosages (*Rao et al., 1988*)

Table 4.2 DEC-values of coagulation/flocculation process in a jar-test for phages and viruses under different conditions (*Nasser et al., 1995*)

Jar-test; 30 mg/l $Al_2(SO_4).18H_2O$ and variable conditions		DEC (log)		
Humic-acid (NOM)	Coagulant aid (cationic)	MS2-phages	Poliovirus	Hepatitis A
–	–	0.3	0.3	0.9
–	+	1.3	0.2	1.8
+	–	0.0	0.3	0.4
+	+	0.4	0.4	0.6

Also in the literature on *Cryptosporidium* and *Giardia* removal by coagulation and flocculation effects of chemical dosage on removal efficiency has been demonstrated. In a pilot plant with sedimentation Dugan *et al.* (2001) showed that at low coagulant dosages (1–10 mg Al/l) the removal of *Cryptosporidium* as well as bacterial spores (aerobic), particles and turbidity was significantly less than the removal of these parameters at coagulant dosages at 10–50 mg Al/l (Figure 4.7).

Results of the same study showed that at the same chemical dosage oocyst removal was higher using Fe-coagulant than Al-coagulant (Table 4.3).

Indications of a better *Cryptosporidium* removal with Fe-coagulant than with Al-coagulant was also seen by Patania *et al.* (1995; Figure 4.8). These data are less supportive for this conclusion, however, because the applied process conditions differed for both coagulants. In the same tests they showed a remarkable difference between the removal of *Giardia* and *Cryptosporidium* (Figure 4.8).

4.4.1.2 Water quality

Nasser *et al.* (1995) demonstrated a negative effect of Natural Organic Matter (NOM; humic acid) on DEC for MS2 phage, poliovirus and Hepatitis A virus (Table 4.2).

The amount of suspended solids in the raw water is also a factor which affects the removal efficiency of coagulation/flocculation. For the removal of *Cryptosporidium* oocysts this was demonstrated by

Patania *et al.* (1995; Figure 4.9). The effect was not observed for *Giardia* cysts. Indirectly the effect of high concentrations of suspended solids was also shown by Cornwell *et al.* (2001). They increased the recycling percentage of back wash water to the influent of the coagulation process and observed an increasing oocyst removal (Figure 4.10). These results must be treated with care, however, since the recovery of the used counting method is low and variable and is influenced by the turbidity of the water.

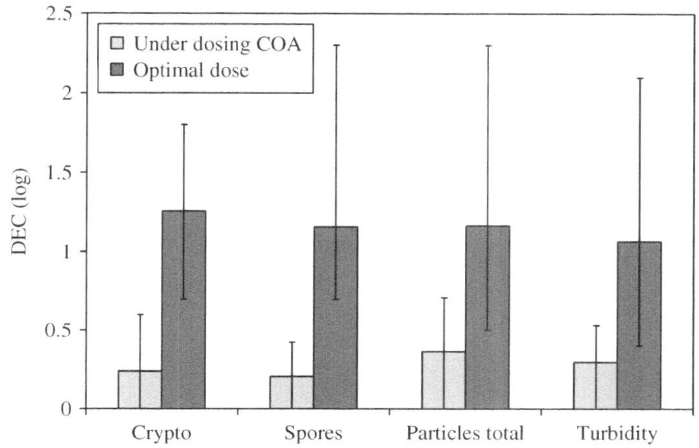

Figure 4.7 DEC of coagulation and sedimentation in a pilot plant at low and high coagulant dosages (*Dugan et al., 2001*)

Table 4.3 DEC of a pilot plant coagulation/sedimentation with either Fe or Al as the coagulant

Concentration	Al	Fe
4–5 mg/l	0.3	0.6
15 mg/l	0.6;0.7	1.2;1.3

4.4.1.3 Flocculation

Generally, data indicated that coagulation followed by sedimentation or lamellae separators is less effective in removing micro-organisms than flocculation with dissolved air (DAF), a conclusion also drawn by Haas *et al.* (2001) and from a large Dutch multiyear study on indicator removal by full-scale *treatment* plants (Table 4.4). DAF or sludge blanket clarifiers (SBC) eliminate SSRC 0.4 log better than LAM under similar influent conditions (turbidity).

Furthermore, coagulation using aluminium and flocculation performed with lamellae separation (LAM) showed a higher elimination capacity for both Coli44 and SSRC than LAM with iron as coagulant on another location (Table 3.5). The difference in coagulant could be the cause of this difference in DEC, but it is more likely that the high turbidity of the source water, especially in winter, was responsible for a higher efficiency of the process with aluminium. High concentrations of suspended solids in the water enhance the coagulation process as discussed before. LeChevallier *et al.* (1991) reported no difference in the efficacy of coagulation with aluminium or iron, but concluded that this could be attributed to the variation in removal by the processes operated under full-scale conditions.

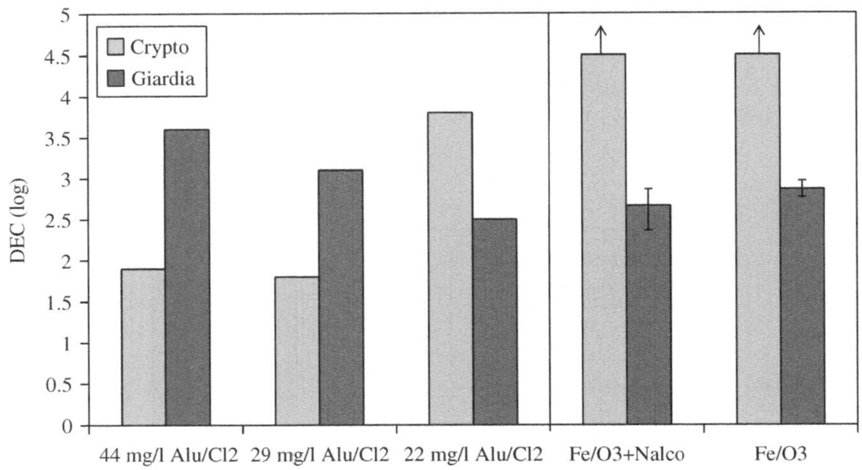

Figure 4.8 DEC of coagulation/flocculation (sedimentation) at different process conditions for protozoan (oo)cysts (*Patania et al., 1995*)

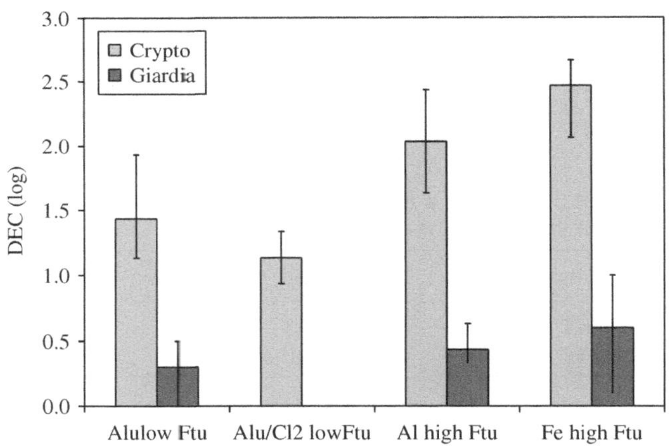

Figure 4.9 DEC of coagulation/flocculation (sedimentation) at different process conditions for protozoan (oo)cysts (*Patania et al., 1995*)

4.4.1.4 Multi regression analysis

Haas *et al.* (2001) used the operational data of studies as input for multiple linear regression analysis and calculated that the DEC of sedimentation was positively correlated with the pH of the influent and the concentration of coagulant and coagulant aid (polymer). For DAF they found a model which showed a correlation with the pH, the recirculation-ratio, the turbidity and the flocculation time.

In a similar study faecal indicator removal data from a full-scale coagulation and flocculation process with LAM were matched with the registered process conditions like lamellae surface load, coagulant concentration, turbidity, coagulant-aid dosage and temperature (Wiersema, 1997). Multi-regression analysis of these data revealed that the removal was negatively correlated with water temperature and the turbidity of the coagulated water (Table 4.5) with low regression coefficients.

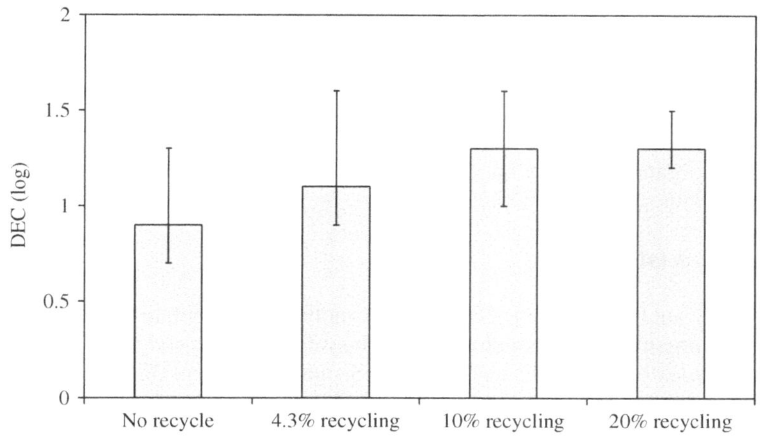

Figure 4.10 DEC of coagulation/flocculation (sedimentation) at different recycle percentages of back wash water for Cryptosporidium (*Cornwell et al., 2001*)

Table 4.4 Elimination of Coli44 and SSRC in coagulation and flocculation by lamellae separation (LAM), dissolved air flotation (DAF) or sludge blanket clarifier (SBC) (*from: Hijnen et al., 2002*)

Flocculation; (Coagulant)[a]	Influent		DEC (log)[c]	
	Turb. (Ftu)	SSRC (n/l)	Coli44	SSRC
LAM (6.5 mg/l Al)	16;6,5[b]	1100; 540[b]	1.3	1.6
LAM (5.0 mg/l Fe)	1.2	11	0.3	0.8
DAF (4.0 mg/l Al)	1.5	33	nd	1.2
SBC (6.0 mg/l Fe)	0.5	67	nd	1.2

[a]Al and Fe is aluminium and iron, respectively; [b]Turbidity and SSRC in winter and summer period, respectively; [c]Data also used for this study; nd = not determined

Table 4.5 The correlation coefficient and level of significance (*p*-value) from a multi-regression analysis of the DEC-values for Coli44 and SSRC removal by coagulation/flocculation with LAM and some routinely monitored process conditions (*from Wiersema, 1999*)

Parameter	Addition of coagulant-aid	Regression coefficient	*P*-values		
			Log C_{in}	Turbidity effluent	Temperature
Coliforms	+	0.26	0.0299	nc	0.257
	−	0.2	0.0068	0.068	nc
SSRC	+	0.35	0.0067	0.04	<0.001
	−	0.5	<0.001	0.0081	<0.001

No correlation

The negative correlation for temperature was due to the use of coagulant-aid in winter. A high removal of micro-organisms at a low effluent turbidity showed that micro-organism removal was related to the removal of suspended solids. Moreover, this suggested that the observed positive correlation between log-transformed SSRC concentrations in the influent and the elimination ($r^2 = 0.50$; $P < 0.001$; $n = 86$)

was caused by a positive correlation between the turbidity and the SSRC concentration in the influent (no turbidity data available to verify).

These data showed that with intensive monitoring of the removal of micro-organisms by full-scale coagulation/flocculation more information can be collected on the relationship between process parameters and elimination. This is of importance for development of on-line process control strategy (Water Safety Plans). It should be emphasised however, that such monitoring programmes should be carefully planned and organised to obtain a time-related and paired data-base and to avoid that some relevant process conditions cannot be evaluated.

4.4.2 FS-index and variation

Based on the sensitivity analysis of the used weighing method to calculate MEC presented in Paragraph 3.3.2 (conventional treatment) it was concluded that the effect of this weighing strongly depends on the distribution of DEC-values of the studies with high FS-indexes. When DEC-values of these studies are equally distributed over the range of values the impact will be small. This is the case for bacteria, bacterial spores and the protozoan (oo)cysts (see black bars in Figures 4.2, 4.3, 4.4 and 4.5). For the viruses more of the studies with high FS-index (black bars) are located at the high range of DEC-values.

In Figure 4.11 the frequency distribution of the DEC-values is presented. A broad distribution of DEC-values was observed for the viruses and the protozoan (oo)cysts, while for bacteria and bacterial spores a more restricted distribution was observed.

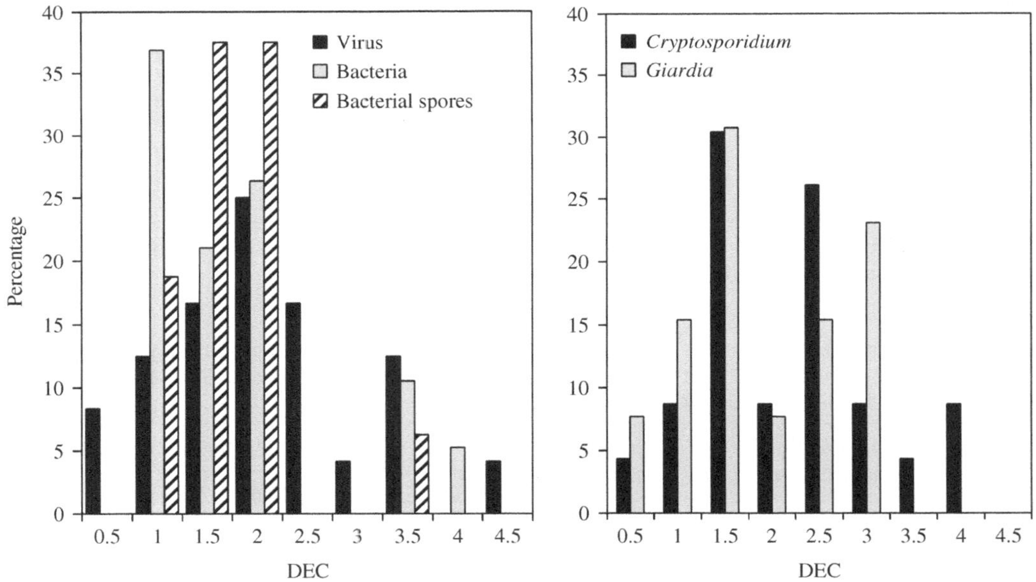

Figure 4.11 The frequency distribution of the DEC-values used to calculate the MEC-values of Table 4.1

The large differences observed in the DEC-values (Figures 4.1–4.5) of the studies were the result of the differences in operational conditions of the studied processes. The applied floc-removal, the coagulant type and dosages, the quality of the incoming water and the hydraulics of the installations all affect the elimination capacity of these processes. Hence, the MEC-values can only be used as an indicative value for the efficacy of coagulation/flocculation.

4.4.3 Surrogates for coagulation/flocculation

The studies of Havelaar *et al.* (1995), Nasser *et al.* (1995) and Bell *et al.* (2000) showed that DEC of the coagulation processes for viruses and phages may vary significantly (Table 4.6). Nonetheless, most of these data indicate that F-spec.-RNA phages can safely be used as surrogates for the quantification of virus removal.

Table 4.6 Variation in DEC observed for different phages and viruses in the dataset

References	Virus	Bacteriophages
Havelaar *et al.*, 1995	1.9 enterovirus; 2.3 Reovirus	1.1 F-spec.RNA
Nasser *et al.*, 1995	0.3 Poliovirus; 1.8 Hepatitis A	1.3 MS2
Bell *et al.*, 2000	3.0 Poliovirus; 1.7 Echovirus	3.1 MS2; 1.5 PRD1

Coagulation and flocculation are water treatment processes with a relative high efficacy for the elimination of viruses, bacteria and bacterial spores and protozoan (oo)cysts. Average MEC of this process ranged from 1.4 up to 1.9 log (Table 4.1). Coagulation and flocculation is most effective for *Cryptosporidium* and viruses followed by bacteria and *Giardia*; the lowest MEC has been calculated for the bacterial spores (1.4 log). It should be emphasised however that this conclusion is based on MEC-values calculated from separate studies. In a number of studies the DEC of coagulation/flocculation was assessed simultaneously for the different micro-organisms (Table 4.7). These data revealed a similar conclusion: DEC-values for viruses, bacteria and protozoa were in the same order of magnitude.

Table 4.7 DEC-values of coagulation processes for the different micro-organisms simultaneously determined (*not all data used for the MEC-values*)

Study	Cryptosporidium	Giardia	Viruses	Bacteria	Bacterial spores
Payment *et al.*, 1993	2.3	3.2	2.3	2.6	2.1
Bell *et al.*, 2000	2.1	2.1	2.3	1.9	1.8
Payment *et al.*, 1985	nd	nd	2	3.3	Nd
Havelaar *et al.*, 1995	nd	nd	0.6	0.6	0.8
Havelaar *et al.*, 1995	nd	nd	1.9	1.7	1.2

Nd = not determined

The relationship between the removal of *Cryptosporidium* oocysts and turbidity, particle count and aerobic spores by coagulation/sedimentation in a pilot plant was determined by Dugan *et al.* (2001). Under optimal and suboptimal coagulation conditions they observed a significant correlation ($P < 0.05$) between the DEC-values of oocysts with all three surrogates (Figure 4.12; $r^2 = 0.88$, 0.88 and 0.86 for turbidity, particles and spores, respectively).

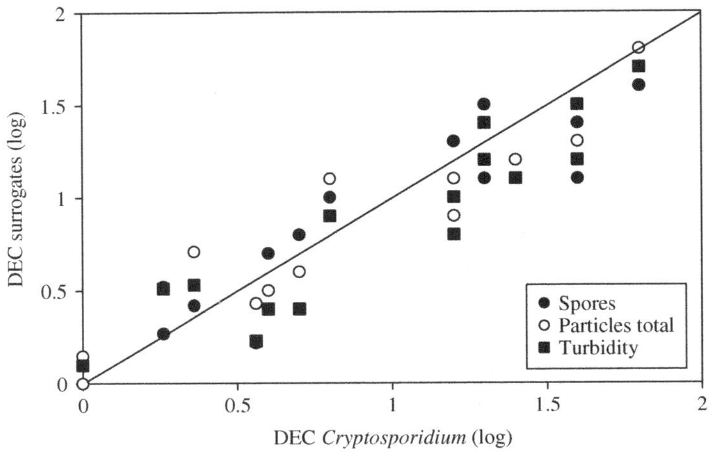

Figure 4.12 DEC of coagulation/sedimentation for oocysts and related surrogates (*Dugan et al., 2001*)

4.5 Conclusions and further research

Although probably not all published literature is included in this evaluation, it covers all references used by Haas *et al.* (2001) and LeChevallier (2001). The evaluation of coagulation/flocculation processes showed that this process is an important barrier in treatment to reduce the concentration of micro-organisms (viruses, bacteria and bacterial spores and (oo)cysts of *Cryptosporidium* and *Giardia*).

- The calculated MEC for the different organisms are in the same order of magnitude and ranged between 1.4 and 1.9 log;
- The FS-index of the evaluated data on removal of bacterial spores was high and moderate for respectively bacteria, viruses and protozoan (oo)cysts;
- Since the process performances for the different groups of micro-organisms were in the same order of magnitude which show that bacteriophages, indicator bacteria and bacterial spores can be used as microbial surrogates for pathogenic viruses, bacteria and protozoa. Turbidity and particle count are physical surrogates which are correlated with *Cryptosporidium* oocyst removal.

The following effects of process conditions on DEC have been observed in literature:

- High turbidity of the influent and pre-oxidation enhance the efficacy of the process;
- Type of flocculation affects the DEC. Data indicate that dissolved air flotation (DAF) and lamellae separation (LAM) removed more organisms than sedimentation (SED);
- Coagulant and coagulant-aid concentrations, pH, temperature are process parameters which have been correlated with the DEC. More specific research on full-scale is needed to decide which process parameters have the largest impact on the variation in treatment efficacy. The results of such research can be used to control the process performance continuously.

5 Rapid granular filtration

Rapid granular filtration (RGF) is widely used in water treatment, especially in the production of drinking water but also as a third or fourth step in the biological sewage water treatment. RGF generally consists of a fixed filter bed of granular media which is normally operated down flow and frequently backwashed to remove accumulated solids and restore hydraulic conditions for water production. The major objectives of RGF are the removal of suspended solids by deep bed filtration and the removal of dissolved substances like iron, manganese, ammonia and organic carbon by biological and physico/chemical processes. Two decades ago the significance of the process for the microbial safety of the produced water was considered to be low. After the large outbreak of *Cryptosporidiosis* in Milwaukee and the increasing knowledge on outbreaks caused by other more persistent pathogenic micro-organisms like *Giardia*, interest in the microbial removal capacity of RGF has increased.

RGF with sand as granular material is applied under three different conditions, possibly influencing the Microbial Elimination Capacity (MEC): a filter bed without a preceding coagulation (RGF) or a filter bed in combination with a preceding coagulation/flocculation, either as a secondary floc-removal process (conventional treatment; RGFcoa) or with in-line coagulation (RGFicoa). In the case of a preceding coagulation/floc-removal a preceding oxidation with chlorine or ozone can be applied. RGF with activated carbon as granular material (RGFgac), has as main objective the removal of organic micro-pollutants. In this chapter the MEC of these processes is calculated for viruses, bacteria, bacterial spores, *Cryptosporidium* and *Giardia*. Both waste water and drinking water studies (WW/DW) were collected to obtain a substantial amount of data for all organisms. This aspect will discussed as part of the variables affecting the DEC-value used for MEC-calucations. In future when DW-literature is extended WW-studies can be omitted from the data base.

5.1 Viruses

RGF (sand/GAC)

Most information on the virus removal by rapid granular filtration available in literature was observed for filter beds without preceding coagulation (RGF; Table 5.1).

The number of drinking water studies (DW) was higher than the number of wastewater studies (WW) and the agreement with full-scale conditions was moderate (FS-index approximately 3). MEC of RGF was 0.8 log (Table 5.1) and the range of DEC-values was high (Figure 5.1). The distribution of DEC-values for WW- and DW-studies (Figure 5.1) was similar indicating that the removal was in the same order of magnitude under these conditions. The high DEC-value of 3.8 log observed in a WW study, however, was largely out of range with the other values. This high value was calculated for the removal of environmental coliphages by a RGF loaded with SW and operated as a slow sand filter (i.e. filtration rate of 0.6 m/h and a contact time of 130 min. (Rose *et al.*, 1996)). Both environmental enteroviruses as well as dosed MS2 coliphages were removed to a lower extent in the same filter; the DEC-values resemble the DEC-values calculated for the other RGF processes

(Figure 5.1). Rose *et al.* (1996) argued that the high DEC for environmental coliphages of 3.8 log was due to the heterogeneity of the group of coliphages containing complex tailed phages and simple isohedral shaped phages.

Table 5.1 MEC of rapid granular filtration for viruses

Conditions	Characteristics				MEC (log)		
	Studies	WW/DW[s]	Data	FS	Average[t]	P50	Range
RGF	13[a,b,c,d,f,h,j,k,l,m,o,p,q]	5/8	90	3.0	0.8 (±0.8)	0.6	0.1–3.8
RGFcoa	5[c,e,g,h,i]	0/5	33	3.1	1.1 (±0.7)	0.8	0.2–2.5
All RGF	18	5/13	123	3.2	0.9 (±0.8)	0.6	0.0–3.8
RGFicoa	6[a,j,n,f,o,p]	2/4	90	3.0	0.9 (±0.5)	0.9	0.1–1.5
RGFgac	4[c,l,r]	0/4	14	2.5	0.4 (±0.2)	nd[u]	0.0–0.7

[a]Robeck *et al.*, 1962; [b]Nestor and Costin, 1971; [c]Guy and McIver, 1977; [d]Sriramula and Chaudhurri, 1982; [e]Payment *et al.*, 1985; [f]De Leon *et al.*, 1986; [g]Rao *et al.*, 1988; [h]Denny and Broberg, 1992; [i]Payment and Franco, 1993; [j]Nasser *et al.*, 1995; [k]Rose *et al.*, 1996; [l]Scott *et al.*, 2002; [m]Watercare 2002; [n]Gitis *et al.*, 2002; [o]Rajala *et al.*, 2003; [p]Scott *et al.*, 2003; [q]Persson *et al.*, 2005; [r]Hijnen *et al.*, 2010; [s]ratio between wastewater WW/drinking water DW studies; [t](±...) Standard deviation; [u]nd = not determined (low number of data)

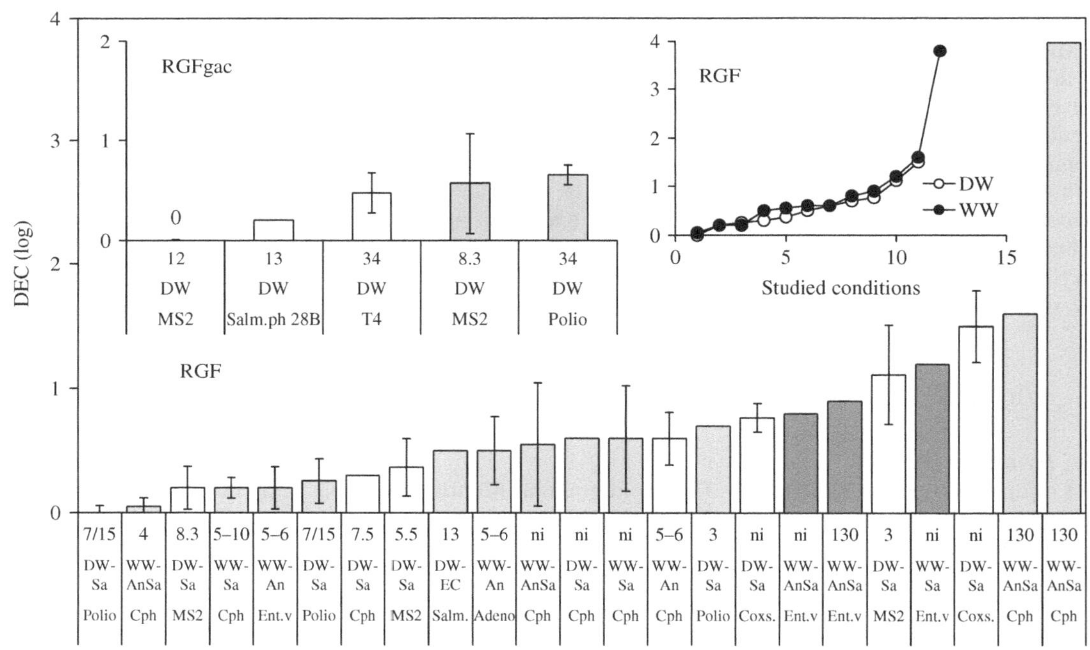

Figure 5.1 The DEC-values (±SD error bars) of the RGF and RGFgac processes observed in literature for viruses and phages operated under different process conditions (X-axis titles from top to bottom: ni = no information; first line = contact time in minutes; second line = Water: WW or DW = waste or drinking water; third line: Filter materials An-Sa = anthracite, sand, EC = expanded clay; fourth line = organisms: Rota = rotavirus, Cph = environmental coliphages, Ent. v = environmental enteroviruses, Polio = poliovirus, MS2 = lab cultured FRNA coliphage, Sam. = Salmonella phage, HAV = hepatitis a virus, Salm.28B, T4 = different kind of phages)

Hardly any elimination of dosed polioviruses was observed by Robeck *et al.* (1962; DW-study) in a RGF with sand (0.78 mm) and a relatively short contact time of 6–12 minutes. The pH ranged from 6–7.5 and temperature was relatively high (15–26°C).

In some studies a variation in DEC was determined (Figure 5.1) and this variation resembles the variation in the average DEC-values (MEC-value) calculated for the accumulated data. The relative standard deviation (SD/Avg) for these processes was 63.2%, ranging from 15.1% to 141%, which was comparable to the relative standard deviation calculated for the MEC-values of 71.3% (omitting the high value of 3.8 log). From this observation it is hypothesised that the variation in DEC of the different studies is not (only) caused by the differences in conditions of these separate studies but (also) by the accumulated variations in conditions of the processes, micro-organisms and analytical methods.

Besides the high DEC-values presented by Rose *et al.* (1996) in a filter bed with high contact time, no other correlations between DEC and either the RGF conditions (water type, granular material) or the studied micro-organisms (pathogens or phages, environmental or dosed) was observed (Figure 5.1). The median value or 50-percentile of the DEC-values is 0.6 log. Based on the collected data a MEC of 0.8 log was calculated for RGF (Table 5.1).

For virus removal by granular activated carbon filtration, RGFgac, less studies were available and these studies were only conducted under drinking water conditions. From these data a MEC-value of 0.4 log was calculated (Table 5.1).

RGF after coagulation

Rapid granular filtration used to remove residual coagulant and poly-electrolyte after sedimentation (RGFcoa), is most likely more effective in removing viruses than RGF without preceding coagulation. Viruses are negatively charged and will attach to positively charged iron or aluminium flocs formed during coagulation or in the filter bed (RGFicoa). From the evaluated studies a MEC was calculated for RGFcoa (*filtration process only without the additional effect of coagulation*) of 1.1 log, which was 0.3 log higher than the MEC calculated for RGF (Table 5.1). This difference, however, was smaller than the variation in the observed DEC-values (Table 5.1) and therefore not significant. Also for (RGFcoa) the variation in DEC-values was high (Figure 5.2) and no correlation between DEC and any of the reported process conditions was found. Payment *et al.* (1985) and Payment and Franco (1993) studied the removal of environmental phages and enteroviruses by anthracite/sand filters operated under full-scale drinking water conditions in combination with a coagulation process. During periods of 1 year concentrations of these micro-organisms were determined by large volume sampling. From these results a MEC of 1.5 log (n = 12; range 1.0–1.9 log) was calculated. This value is considered as reference value for the MEC-values calculated from the accumulated data set and show that RGF in combination with a preceding coagulation is at least 0.5 log more effective in the removal of viruses than a single RGF.

In line coagulation

As observed for filtration after coagulation and floc-removal, a positive effect of in-line coagulation on MEC was expected. From the results presented in Table 5.1, however, no significant increase in MEC was observed for these filters (RGFicoa) compared to RGF. This indicates that the coagulation process occurring in the filter bed did not enhance virus removal. The high range in DEC-values observed in some of these studies (Figure 5.2) suggests that MEC of these filters is more variable, although this was calculated from the overall data base (SD = 0.5). RGFicoa filled with sand were more efficient in virus removal than filters filled with anthracite/sand. All studies evaluated used aluminium as the coagulant and from the presented data no effect of the coagulant dose was observed (Figure 5.2).

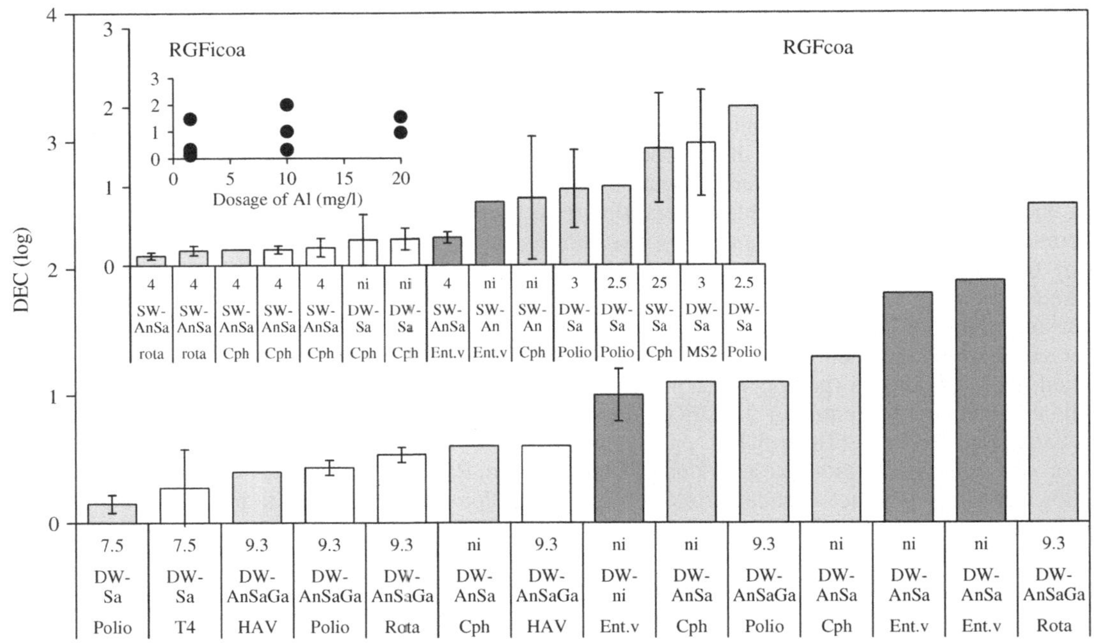

Figure 5.2 The DEC-values (±SD error bars) of the RGFcoa and RGFicoa processes observed in literature for viruses and phages operated under different process conditions (X-axis titles from top to bottom: ni = no information, first line = contact time in minutes; second line = Water: WW or DW = waste or drinking water; third line: Filter materials AnSaGa = anthracite, sand, garnet; fourth line = organisms: Rota = rotavirus, Cph = environmental coliphages, Ent. v = environmental enteroviruses, Polio = poliovirus, MS2 = lab cultured FRNA coliphage, Sam. = Salmonella phage, HAV = hepatitis a virus)

5.2 Bacteria and bacterial spores

RGF (sand/GAC; bacteria)

The number of studies describing bacteria removal by RGF was limited (Table 5.2) and the FS-index was relatively low. The majority of these studies (11) looked into the effect of RGF with or without a preceding coagulation.

Table 5.2 MEC of rapid sand filtration for bacteria

Conditions	Characteristics				MEC (log)		
	Studies	WW/DW	Data	FS	Average	P50	Range
RGF	6[b,e,h,i,k,l]	2/4	453	2.8	0.5 (±0.3)	0.4	0.2–1.0
RGFcoa	5[c,d,e,f,g]	1/4	60	3.0	0.9 (±0.4)	0.9	0.4–1.5
All RGF	11	3/8	513	2.9	0.7 (±0.4)	0.6	0.2–1.5
RGFicoa	4[a,l,m,n]	3/1	31	3.0	1.4 (±0.4)	1.4	0.8–2.1
RGFgac	4[f,h,j,o]	0/4	19	3.0	0.7 (±0.3)	0.8	0.1–0.9

[a]Cleasby *et al.*, 1984; [b]Haas *et al.*, 1985; [c]Payment *et al.*, 1985; [d]Nieuwstad *et al.*, 1988; [e]Denny *et al.*, 1992; [f]Gray *et al.*, 1999; [g]Hijnen *et al.* 1997a; [h,i,j]Hijnen *et al.*, 1998b,d,e; [k]Watercare, 2002; [l]Rajala *et al.*, 2003; [m]Scott *et al.*, 2003; [n]Koivunen *et al.*, 2003; [o]Hijnen *et al.*, 2010

Only one study monitored the removal of a pathogenic bacterium, *Campylobacter* (Hijnen *et al.*, 1998d; Figure 5.3) by full-scale RGF; the number of samples collected was 29. The DEC of the RGF process was 0.5 (±0.6) log. The simultaneously monitored DEC of the process for thermotolerant coliforms (Coli44) was 0.7 (±0.4) log, indicating that application of these indicator bacteria as surrogates for *Campylobacter* removal by RGF needs a correction factor of 0.7 (ratio between DEC-values).

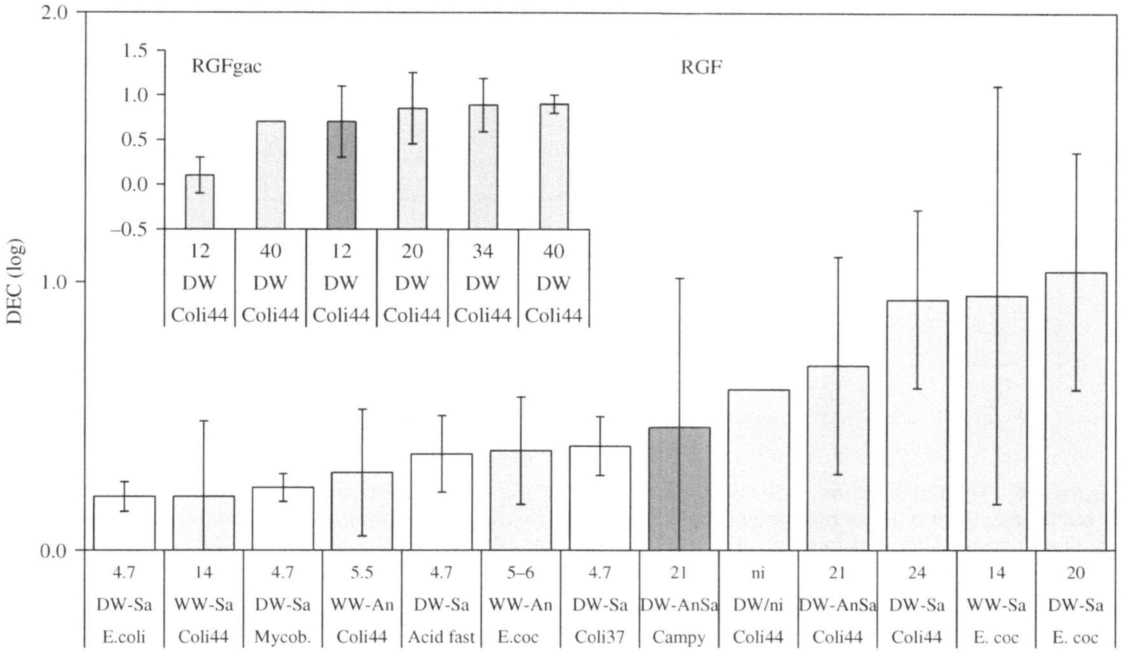

Figure 5.3 The DEC-values (±SD error bars) of the RGF and RGFgac processes observed in literature for bacteria operated under different process conditions (X-axis titles from top to bottom: ni = no information, first line = contact time in minutes; second line = Water: WW or DW = waste or drinking water and filter materials An-Sa = anthracite, sand; third line = organisms, Coli44/Coli37 = thermotolerant and total coliforms; Mycob. = Mycobacterium; Acidfast = yeast; Campy = Campylobacter; E. coc = enterococci

In two wastewater studies a relatively low DEC (0.2 and 0.4 log) was observed for Coli44 (faecal coliforms; Figure 5.3). The data showed a correlation between DEC and the hydraulic contact time with higher removal in the RGF processes at higher contact times. Most likely the high contact time in the RGF_{GAC} of 20–40 minutes (Figure 5.3) caused the higher MEC-value of 0.7 log calculated for these filters (Table 5.2).

RGF after coagulation and with in line coagulation (bacteria)

With a preceding coagulation/sedimentation process (RGFcoa) the DEC of these filters increased to a MEC-value of 0.9 log and with in-line coagulation (RGFicoa) the removal efficiency increased again with 0.5 log to 1.4 log (Table 5.2; Figure 5.4). The two RGFcoa filters tested with waste water showed low DEC-values. Because of lack of information, for these filters a relationship with the contact time could not be observed. For the in-line filters RGFicoa the higher DEC-values were observed for filters loaded with waste water and higher contact times (undecided which condition had the highest impact). Studies on the removal of bacteria by full-scale filters with coagulation was not found in literature.

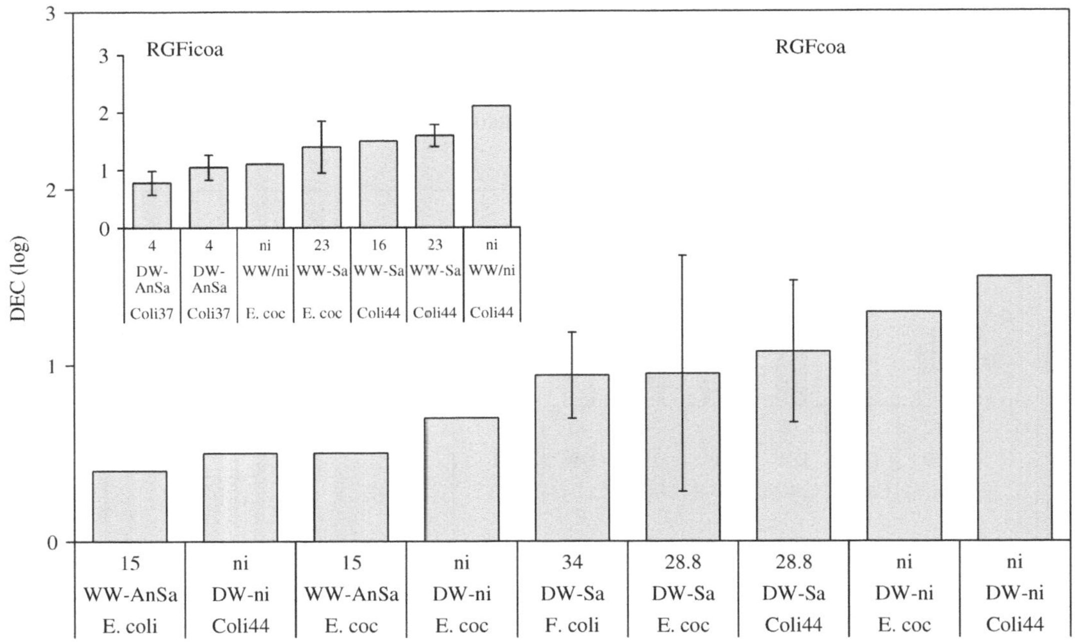

Figure 5.4 The DEC-values (±SD error bars) of the RGF_{coa} and RGFicoa processes observed in literature for bacteria operated under different process conditions (X-axis titles from top to bottom: ni = no information, first line = contact time in minutes; second line = Water: WW or DW = waste or drinking water and filter materials An-Sa = anthracite, sand; third line = Organisms, Coli44/Coli37 = thermotolerant and total coliforms; E.coc = enterococci)

Bacterial spores removal: RGF (sand/GAC)

Compared to bacteria, bacterial spores are removed more efficiently by RGF processes (Table 5.3), especially with a preceding coagulation or in-line coagulation. Only few studies were found on the removal of spores during waste water treatment. The MEC-value of RGF was 0.7 log with a relatively large variation (SD = 0.5 log; range = 0–1.4 log). Looking into the details of the high variation showed that the anaerobic spores (SSRC or spores of *C. perfringens*) were removed to a higher degree than the aerobic spores (ARS or spores of *B. subtilis*) by RGF (Figure 5.5).

Table 5.3 MEC of rapid sand filtration for bacterial spores

Conditions	Characteristics				MEC (log)		
	Studies	WW/DW	Data	FS	Average	P50	Range
RGF	7[a,j,l,n,q,r,s]	2/5	106	4.1	0.7 (±0.5)	0.4	0.0–1.4
RGFcoa	10[a,b,c,e,f,g,k,n,p,q]	0/10	123	4.1	1.6 (±0.8)	0.8	0.5–2.9
All RGF	14	2/12	229	4.1	1.2 (±0.9)	1.2	0.0–2.9
RGFicoa	5[d,h,o,q,s]	1/4	30	3.8	2.3 (±0.7)	2.2	1.5–3.9
RGFgac	8[c,d,e,i,j,k,m,t]	0/8	64	5.0	0.7 (±0.4)	0.5	0.1–1.5

[a]Denny *et al.*, 1992; [b]Payment *et al.*, 1993; [c]Hijnen *et al.*, 1994; [d]Rice *et al.*, 1996; [e]Gray *et al.*, 1999; [f]Hijnen *et al.*, 1997a; [g]Yates, *et al.*, 1997a; [h]Yates *et al.*, 1997b; [i,j,k,l,m] Hijnen *et al.*, 1998abcde; [n]Coffey *et al.*, 1999; [o]Swertfeger *et al.*, 1999; [p]Dugan *et al.*, 2001; [q]Emelko *et al.*, 2001; [r]Watercare, 2002; [s]Rajala *et al.*, 2003; [t]Hijnen *et al.*, 2010

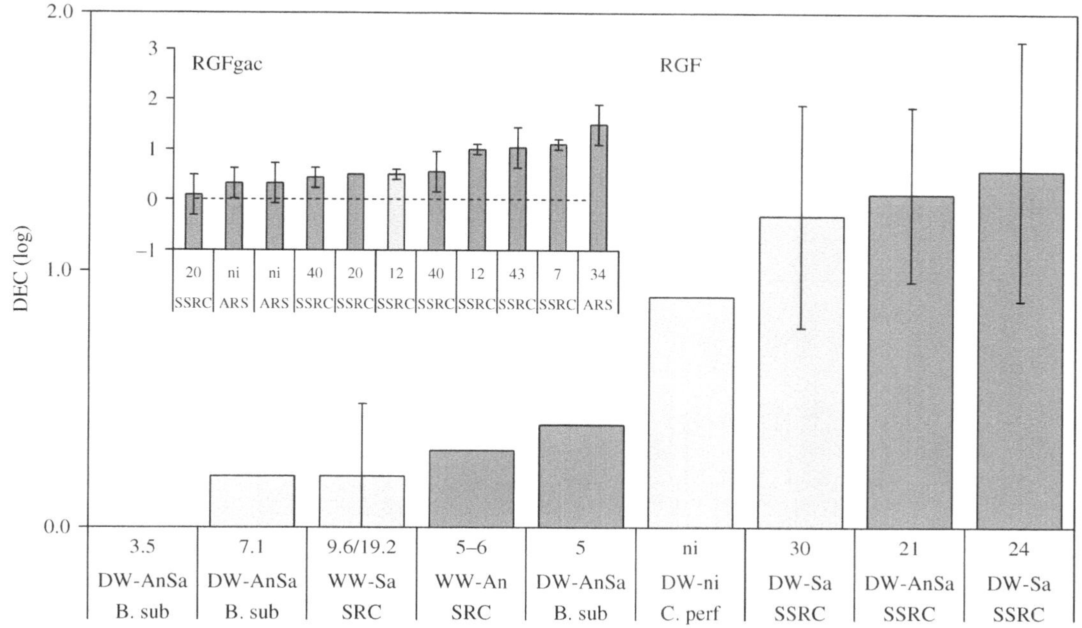

Figure 5.5 The DEC-values (±SD error bars) of the RGF and RGFgac processes observed in literature for bacterial spores operated under different process conditions (X-axis titles from top to bottom:ni = no information, first line = contact time in minutes; second line = Water: WW or DW = waste or drinking water and filter materials An-Sa = anthracite, sand, GAC = granular activated carbon; third line = organisms, ARS = aerobic spore forming bacteria, B. sub. = Bacillus subtilis, SRC = sulphite-reducing clostridia, SSRC = spores of sulphite-reducing clostridia, C. perf. = Clostridium perfringens)

In two studies, one for environmental SSRC removal and another for environmental ARS by GAC-filtration, the relative SD was >100%. In the study of environmental SSRC (Hijnen *et al.*, 1994) negative removal was observed; higher concentrations coming out than coming in). This phenomenon of breakthrough was described first in literature by Hijnen *et al.* (1997b) and attributed to accumulation, survival and delayed breakthrough of spores in these filters with a low backwash intensity and frequency. This delayed breakthrough was also proposed as mechanism responsible for the observed low DEC of slow sand filtration for SSRC (Hijnen *et al.*, 2000). Recently, Mazoua and Chauveheid (2005) described the same mechanism for aerobic spore removal by GAC-filtration although regrowth in the filter bed is another possible explanation for the low DEC or sometimes even negative removal. Regrowth of anaerobic spores is less likely although not completely impossible as described in literature before (Willis, 1957). This author was one of the first to "lay emphasize on the value of the anaerobe test on treated waters in assessing filter efficiency" and he also pointed out the significance of aftergrowth of spores in slow sand filters. The significance of the observed SSRC-aftergrowth in these filters, however, is debatable since filtration conditions were not described. At that time (1957) these filters were uncovered open filters with algae bloom on top of the filter bed. Under these conditions oxygen depletion, especially when operation is intermittent, may occur with growth of SSRC as a result.

The high DEC-values of RGF for spores were observed in sand or anthracite/sand filters with relatively high hydraulic retention times of 21–30 minutes, whereas the RGF processes with lower contact times showed lower DEC-values (Figure 5.5). Furthermore, the efficiency of RGF and RFGgac to remove spores is similar (Table 5.3).

Bacterial spores (RGF after coagulation)

The MEC-value for RGF with a preceding coagulation RGFcoa of 1.6 (±0.8) log was also highly variable (Figure 5.6). A better removal of the anaerobic spores (SSRC, *C. perfringens*) than the removal of aerobic spores (ARS, *B. subtilis*) was not observed for these filters. Furthermore, there was no correlation between DEC and the contact time. Possibly, the process conditions in the coagulation/sedimentation process prior to the filtration rather than the contact time determine the DEC of the subsequent filtration process for the spores. These organisms were evaluated as surrogates for protozoan (oo)cysts but because of their smaller size they are evaluated as a separate group thus giving relatively high FS-values (Table 5.3).

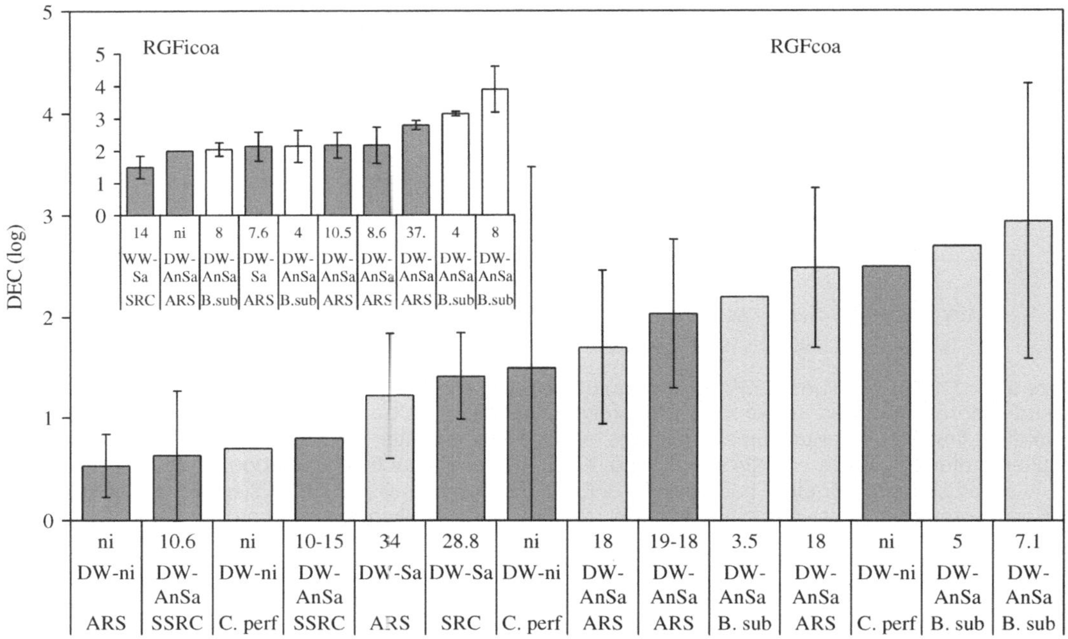

Figure 5.6 The DEC-values (±SD error bars) of the RGFcoa and RGFicoa processes observed in literature for bacterial spores operated under different process conditions (X-axis titles from top to bottom: ni = no information, first line = contact time in minutes; second line = Water: WW or DW = waste or drinking water and filter materials An-Sa = anthracite, sand, GAC = granular activated carbon; third line = organisms, ARS = aerobic spore forming bacteria, B. sub. = Bacillus subtilis, SRC = sulphite-reducing clostridia, SSRC = spores of sulphite-reducing clostridia, *C. perf.* = Clostridium perfringens

5.3 *Cryptosporidium* and *Giardia*

The large waterborne outbreak of *Cryptosporidium* in Milwaukee (MacKenzie *et al.*, 1994) initiated worldwide research on the removal of protozoan (oo)cysts by filtration processes. For the rapid granular filtration processes, most of studies in the research have been carried out with dosage of oocysts rather than monitoring the removal of environmental oocysts. Due to the scale of these studies (pilot plant scale), the average FS-value was moderate. One waste water and one drinking water study were found where removal of environmental oocysts under full-scale filtration conditions was examined (Figure 5.7; dark bars).

The calculated MEC-value for RGF processes without pretreatment is 1.1 (±0.7) log with a range of values of 0–2.3 log (Table 5.4). No clear effect of the contact time and filter media on the DEC could be observed from the data (Figure 5.7). The low DEC-values (0–0.5) were observed by Cofey *et al.* (1999), Emelko *et al.*, 2001, Gitis *et al.* (2002) and Haas *et al.* (1985), Patania *et al.* (1995). Patania observed no removal in a dual media filter bed with GAC and sand with a relatively high filtration rate (15 m/h). The other authors presenting low DEC-values described dosage experiments with anthracite/sand or sand filters with no significantly different operational or design differences as in the other presented studies.

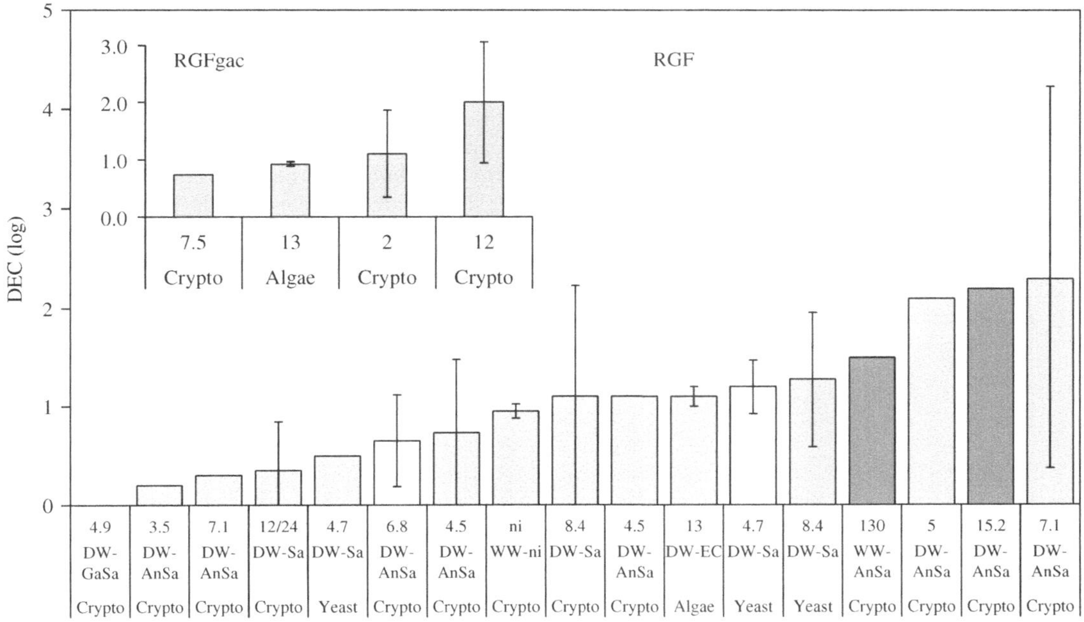

Figure 5.7 The DEC-values (±SD error bars) of the RGF and RGFgac processes observed in literature for oocysts of Cryptosporidium operated under different process conditions (X-axis titles from top to bottom: first line = contact time in minutes; second line = Water: WW or DW = waste or drinking water and filter materials An-Sa = anthracite, sand, GAC = granular activated carbon, EC = expanded clay; third line = organisms)

Table 5.4 MEC of rapid sand filtration for Cryptosporidium

Conditions	Characteristics				MEC (log)		
	Studies	WW/DW	Data	FS	Average	P50	Range
RGF	10[a,g,h,j,o,r,t,v,w,y]	2/8	46	3.1	1.1 (±0.7)	1.1	0.0–2.3
RGFcoa	11[b,d,g,i,j,l,n,p,q,r,s]	0/11	104	3.5	2.6 (±1.3)	2.1	0.8–5.5
All RGF	21	2/19	156	3.3	1.8 (±1.3)	1.7	0.0–5.5
RGFicoa	11[c,e,f,g,k,m,o,r,t,u,v,x]	0/11	238	2.7	3.0 (±1.4)	2.9	0.1–5.4
RGFgac	3[g,y,z]	0/3	12	3.0	1.3 (±0.6)	1.1	0.7–2.7

[a]Haas *et al.*, 1985; [b]Payment and Franco, 1993; [c]Nieminski *et al.*, 1994; [d]Kelley *et al.*, 1995; [e]Nieminski *et al.*, 1995; [f]Ongerth *et al.*, 1995; [g]Patania *et al.*, 1995; [h]Rose *et al.*, 1996; [i]Swaim *et al.*, 1996; [j]States *et al.*, 1997; [k]Swertfeger *et al.*, 1997; [l]Yates *et al.*, 1997a; [m]Yates *et al.*, 1997b; [n]Edzwald *et al.*, 1998; [o]Cofey *et al.*, 1999; [p]Cornwell *et al.*, 2001; [q]Dugan *et al.*, 2001; [r]Emelko *et al.*, 2001; [s]Akiba *et al.*, 2002; [t]Gitis *et al.*, 2002; [u]Kim *et al.*, 2002; [v]Scott *et al.*, 2003; [w]Chung *et al.*, 2004; [x]Dugan *et al.*, 2004; [y]Persson *et al.*, 2005; [z]Hijnen *et al.*, 2010

RGF (sand/GAC; Cryptosporidium)

The two studies on removal of environmental oocysts by full-scale RGF (FS = 5), Rose *et al.* (1996) (waste water) and States *et al.* (1997) (drinking water), revealed DEC-values in the higher range (1.5 and 2.0 log, respectively; Figure 5.7). The few studies examining GAC-filtration showed a large range of DEC-values (0.77–2.7 log) and the MEC-value of this process was slightly higher than the MEC of RGF (Table 5.4 and Figure 5.7).

RGF after coagulation (Cryptosporidium)

A preceding coagulation/floc removal step before filtration (RGFcoa) enhances the removal of oocysts significantly; the MEC increased from 1.1 without pre-coagulation to 2.1–2.5 log on an average. The removal of environmental *Cryptosporidium* by full-scale filters determined by Payment and Franco (1993), Kelley *et al.* (1995) and States *et al.* (1997) was 0.8, 1.8 and 2.2 log, respectively. Furthermore, also the removal of environmental algae (*Cyanophyta, Chlorophyta* and *Diatomaea*) as surrogate for *Cryptosporidium* by full-scale filters monitored by Akiba *et al.*, (2002) revealed DEC-values in the lower range of the DEC-values (Figure 5.8). The high DEC-values (>3 log) were all dosage studies. Based on these considerations it is proposed to use the median value of 2.1 log rather than the arithmetic value of 2.5 log.

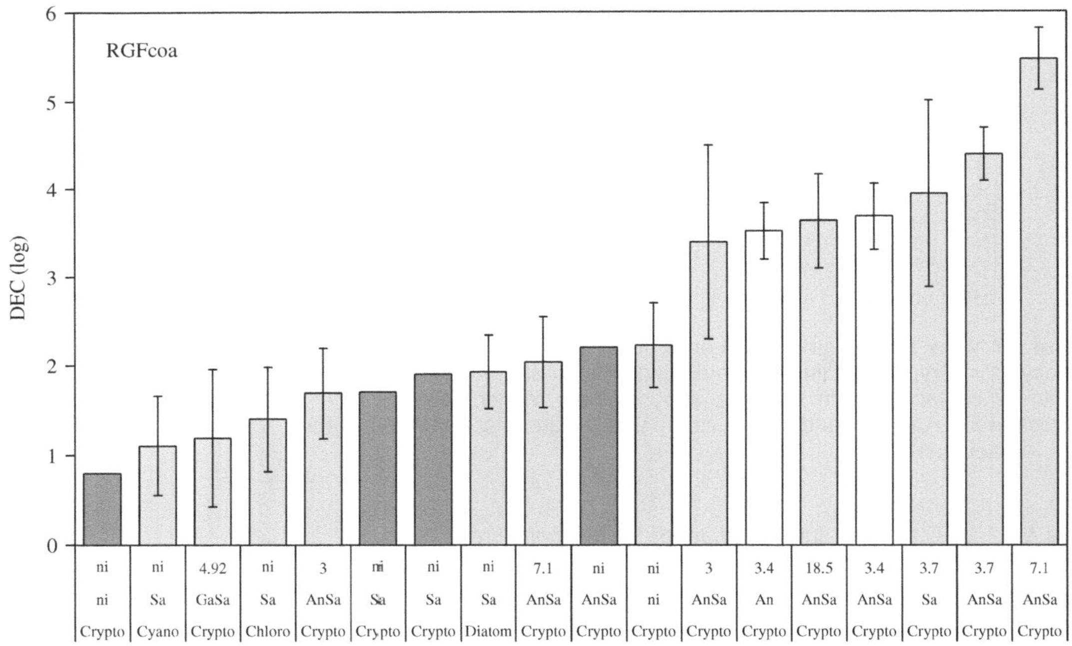

Figure 5.8 The DEC-values (±SD error bars) of the RGFcoa processes observed in literature for oocysts of Cryptosporidium operated under different process conditions (X-axis titles from top to bottom: first line = contact time in minutes; second line = filter materials AnSaor GaSa = anthracite or GAC, sand, GAC = granular activated carbon; Third line = organisms)

RGF with in line coagulation (Cryptosporidium)

Using direct in-line coagulation (RGFicoa) the MEC further increased to 3.0 log (Table 5.4; Figure 5.9). The study of Scott *et al.* (2003) described the removal of environmental oocysts in three wastewater plants; one of these plants used anthracite/sand filters with polymer addition. No operational information about filtration rate and backwashing was presented and hardly any removal of *Cryptosporidium* was observed in these filters.

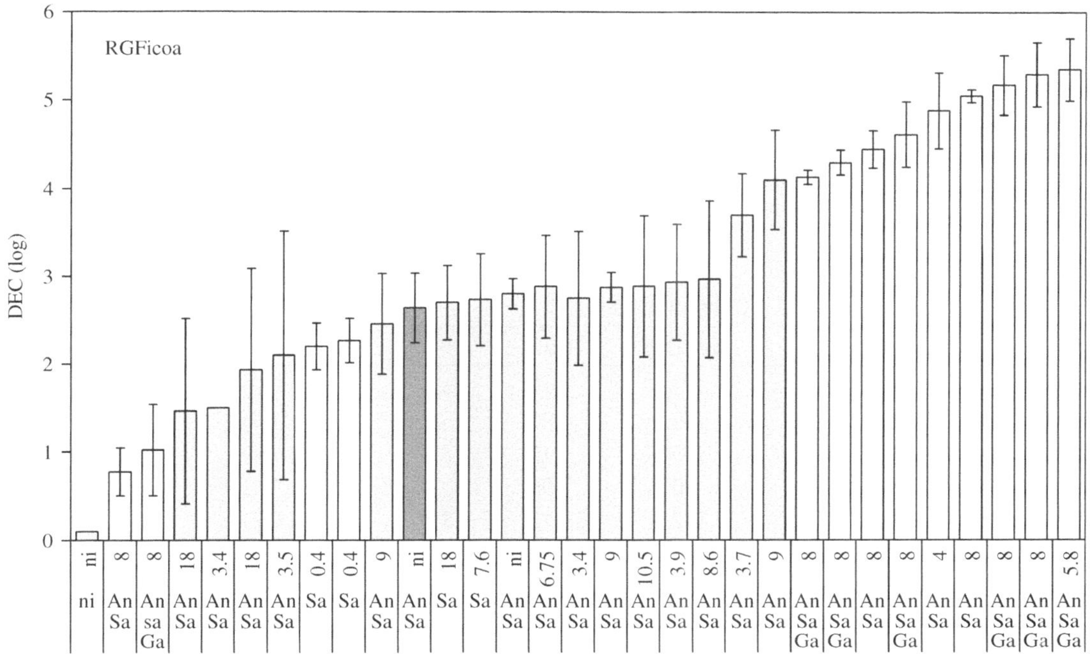

Figure 5.9 The DEC-values (±SD error bars) of the RGFicoa processes observed in literature for oocysts of Cryptosporidium operated under different process conditions (X-axis titles from top to bottom: first line = Contact time in minutes; second line = filter materials AnSaor GaSa = anthracite or GAC, sand, GAC = granular activated carbon)

The other values of <1.5 log were from the study of Emelko *et al.* 2001 and were monitored during coagulation failures. In the discussion in § 5.4 this study will be described in more detail. The DEC-values of >3 log were observed in the studies of Swertfeger *et al.* (1999), Yates *et al.* (1997b), Dugan *et al.* (2004) and the eight highest values came from the study of Emelko *et al.* (2001) with small scale filters. These removal data were obtained in a diverging experimental set up where a suspension of pre-coagulated oocysts on laboratory scale was dosed to the influent of the filter bed.

RGF (sand/GAC; Giardia)

Compared to *Cryptosporidium* the removal of *Giardia* cysts by RGF is calculated from a lower number of studies (Table 5.5) and most of these studies used dosage of cysts to monitor removal (moderate FS). The MEC of RGF and RGFgac for cysts was similar to the MEC calculated for oocysts (Table 5.4). Because Logsdon *et al.* (1985) studied the removal of cysts by GAC-filtration with preceding coagulation and sedimentation, this value of >3 log (Figure 5.10) was not included in the MEC-calculation.

Low removal of *Giardia* (≤0.5) by RGF was observed by Patania *et al.* (1995); DEC of 0 and 0.2 log was observed in filters operated at relatively high filtration rates of 18 and 24 m h^{-1}, respectively with contact time of 4.1 and 5.6 minutes (Figure 5.10). Kelley *et al.* (1995) studied the removal of environmental *Giardia* cysts by full-scale sand filters. The study, however, did not mention the operational data of the filters. The DEC-values from this study were equal or larger than the MEC-value of 1.2 log.

RGF after coagulation and with in line coagulation (Giardia)

Again, preceding coagulation and in-line coagulation increased the MEC-value of RGF considerably (Table 5.5). Payment *et al.* (1993) and States *et al.* (1997) monitored the removal of environmental cysts by RGFcoa (FS = 5; Figure 5.11) and these values were close to the MEC-value calculated for this kind of filters and somewhat higher than the MEC calculated for RGF (Table 5.5). A low DEC-value of <0.5 log was presented by Al-ani *et al.* (1985) in filters with a short contact time of 1.8 minutes (Figure 5.11). The in-line coagulation filters (RGFicoa) showed a range of DEC-values of 0.8–4.7 log (Table 5.5 and Figure 5.12) with a calculated MEC-value of 2.9 log.

Table 5.5 MEC of rapid sand filtration for Giardia

Conditions	Characteristics				MEC (log)		
	Studies	WW/DW	Data	FS	Average	P50	Range
RGF	5[a,g,i,j,n]	2/3	17	3.5	1.2 (±0.9)	1.0	0.0–2.6
RGFcoa	6[b,c,d,i,k,l]	0/6	96	3.2	1.9 (±1.1)	1.7	0.3–4.1
All RGF	11	2/9	113	3.3	1.7 (±1.1)	1.5	0.0–4.1
RGFicoa	7[a,b,e,f,h,i,m,n]	1/6	105	3.0	2.9 (±1.1)	3.0	0.8–4.7
RGFgac	2[i,o]	0/2	12	3.0	1.4 (±0.7)	1.6	0.4–2.0

[a]Logsdon *et al.*, 1981; [b]Al-Ani *et al.*, 1985; [c]Logsdon *et al.*, 1985; [d]Payment and Franco, 1993; [e]Nieminski *et al.*, 1994; [f]Nieminski *et al.*, 1995; [g]Kelley *et al.*, 1995; [h]Ongerth *et al.*, 1995; [i]Patania *et al.*, 1995; [j]Rose *et al.*, 1996; [k]Swaim *et al.*, 1996; [l]States *et al.*, 1997; [m]Swertfeger *et al.*, 1997; [n]Scott *et al.*, 2003; [o]Hijnen *et al.*, 2010

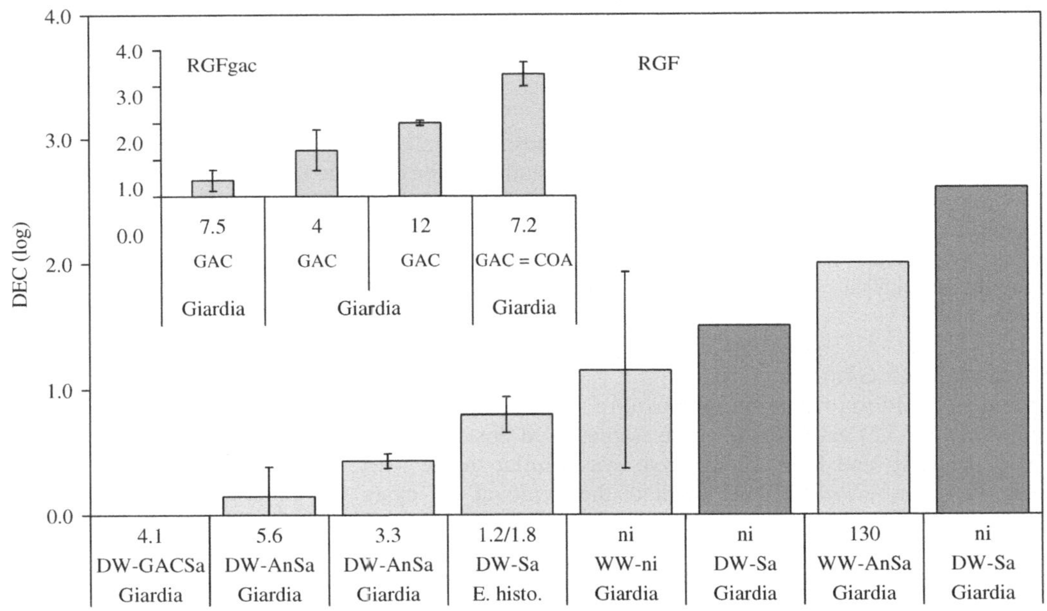

Figure 5.10 The DEC-values (±SD error bars) of the RGF and RGFgac processes observed in literature for cysts of Giardia operated under different process conditions (X-axis titles from top to bottom: first line = contact time in minutes; second line = Water: WW or DW = waste or drinking water and filter materials An-Sa = anthracite, sand, GAC = granular activated carbon; third line = organisms)

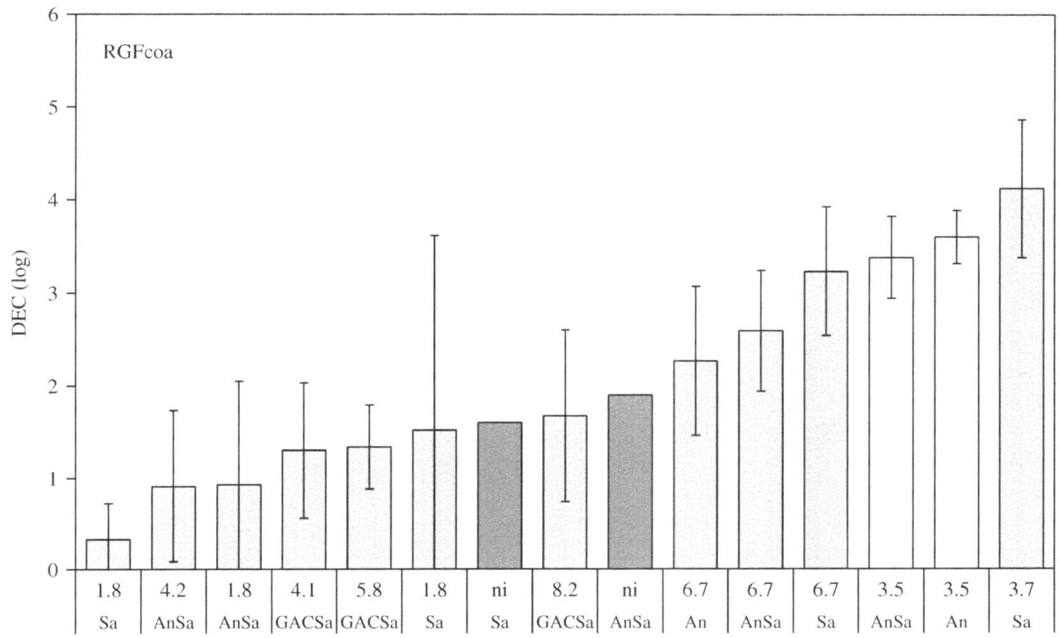

Figure 5.11 The DEC-values (±SD error bars) of the RGF and RGFgac processes observed in literature for cysts of Giardia operated under different process conditions (X-axis titles from top to bottom: first line = contact time in minutes; second line filter materials An-Sa = anthracite, sand, GAC = granular activated carbon)

5.4 Discussion

The results of this review on Microbial Elimination Capacity (MEC) of rapid granular filtration for viruses, bacteria, bacterial spores and protozoan oocysts show a considerable variation in removal efficiency (Figure 5.13). A variation between the different types of filters and organisms and also between studies investigating the same type of organism and filtration process (error bar shows the range of DEC-values). This variation is obviously related to the variation in conditions of the evaluated studies like studied micro-organisms (environmental–pre-cultured), water quality, water temperature, process design and process operational conditions.

The mechanisms affecting removal in granular filtration processes are:

- physical straining;
- attachment/detachment;
- survival: decay or predation.

Since the contact time in RGF is relatively short, it is assumed that factors like decay and predation will have limited effect on micro-organism removal capacity.

The removal mechanisms are influenced by the filtration conditions related to

- the characteristics of the micro-organism and the granular media like size and surface properties;
- the design of the process: with or without pre-coagulation, filtration rates and contact time, granular media (type and grain size);
- operational conditions: temperature, hydraulic steps, backwash strategy, ripening and breakthrough period.

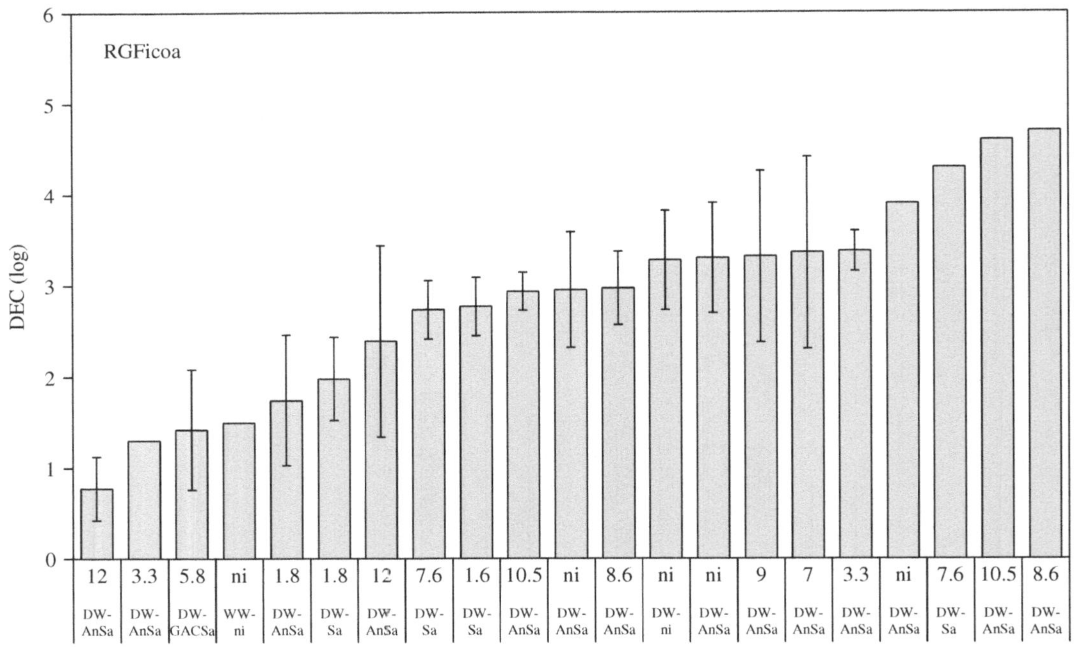

Figure 5.12 The DEC-values (±SD error bars) of the RGFicoa processes observed in literature for cysts of Giardia operated under different process conditions (X-axis titles from top to bottom: first line = contact time in minutes; second line filter materials An-Sa = anthracite, sand, GAC = granular activated carbon)

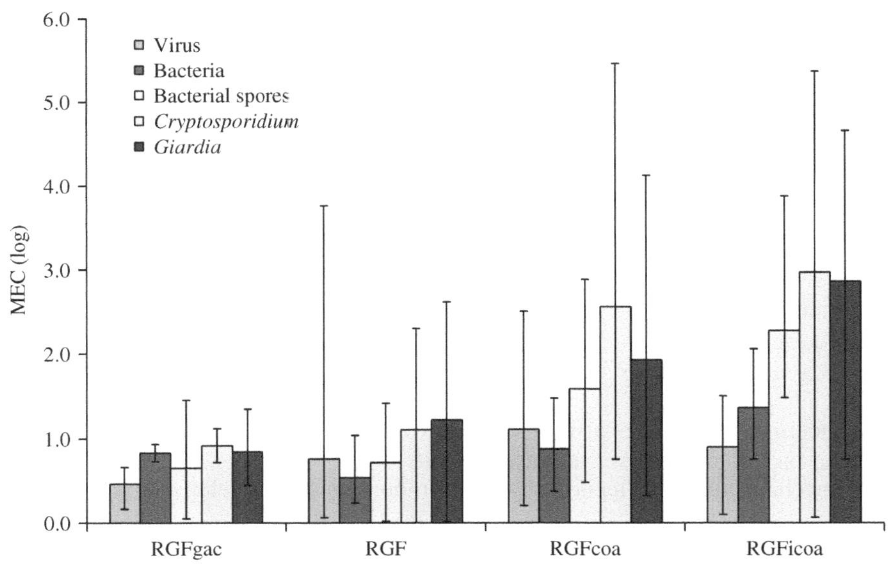

Figure 5.13 The MEC (error bars is range of DEC-values) of the different RGF processes for viruses, bacteria, bacterial spores and protozoan (oo)cysts

To some extent the sequence in MEC of the different filtration types seems to be related to the size of the organism which is an indication that straining attributes to the overall removal. Generally, MEC of the process increased when coagulation is used as a pre-treatment step or as an additional step in the filter bed. The higher MEC-value of RGF and RGFcoa for viruses compared to the MEC of these processes for bacteria suggests a significant role of attachment in the removal. This is supported by observations of different DEC-values for micro-organisms or particles of the same size as presented before. Several studies paid attention to these aspects of filtration and some of these are presented in the following sections (restricted to RGF studies; numerous studies on micro-organism transport in granular materials are present in literature).

5.4.1 Factors related to straining

Straining as a removal mechanism, which will result in irreversible removal, is likely to occur in the small sized pores in the filter bed and is of influence on the removal of bacter, bacterial spores and protozoan (oo)cysts. Rapid granular filters are backwashed frequently to restore porosity which will reduce the pore volumes in the smaller size ranges and therefore limit the contribution of straining to the overall removal most likely. Increased DEC-value during the first period after back washing (ripening) and higher DEC-values observed for filters with preceding coagulation (higher load of suspended solids) is most likely related to increased contributuion of straining; reduction of the porosity during ripening or maturation will most likely increase the percentage of small sized pores. Continuation of the loading of the filter bed leads to a reduction in the percentage of small sized pores and thus a reduction in straining, eventually leading to breakthrough of particles. Both aspects of ripening and breakthrough are discussed in the next sections.

5.4.2 Factors related to attachment

5.4.2.1 Surface properties and electrostatic interactions

Attachment of biocolloids such as micro-organisms to granular media depends on the surface properties like hydrophobicity and electric charge. Hydrophobicity and charge depend on the chemical components present on the surfaces and the surface charge is also influenced by ion concentration and pH of the surrounding water. For virus removal attachment is probably the dominating mechanism. The isoelectric point (zero charge pH) of viruses and other micro-organisms as well as granular materials is usually below pH 6, so then have a negative surface charge at neutral and higher pH-values. Consequently, the attraction of micro-organisms to clean granular materials is hindered by repulsion, which increases at high pH. This was observed to some extent by Foliguet and Doncoeur (1975); they observed a net lower poliovirus removal by coagulation ($FeCl_3$) and RGF at increased pH-values during coagulation (Figure 5.14). In this study, however, two counteracting effects may have resulted in the observed relationship: higher repulsion as described above with decreased filtration efficiency or decreased coagulant accumulation in the filter bed at high pH due to a more efficient preceding coagulation process. A lower coagulant accumulation will have a negative impact on polio removal in the filter bed.

A more direct experiment, though on lab-scale with glass beads, showing the effect of pH on attachment efficiency of protozoa (oo)cysts, was presented more recently by Hsu *et al.* (2001) and also by other authors investigating oocysts transport in soils like Tufenkji and Elimelech (2005; results not presented). The removal efficiency of the filter for (oo)cysts observed by Hsu decreased with pH (Figure 5.14); this effect was more pronounced for *Giardia* than for *Cryptosporidium*, indicating different surface properties.

Also the charge of granular media affects the removal of viruses. From the data of Sriramula and Chaudhuri (1982) and Scott *et al.* (2002) it was observed that removal of MS2 and PRD1, two bacteriophages, by small scale filter columns increased when the filter media were coated with Al or Fe-hydroxides or polyelectrolyte. This coating resulted in an increase of the zeta-potential (more positively charged; Figure 5.14).

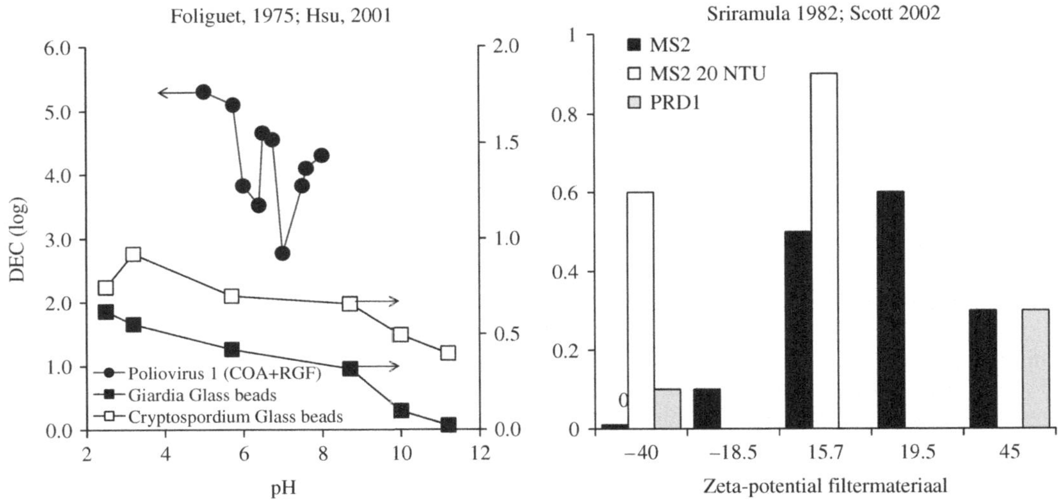

Figure 5.14 The effect of pH of the water (left; notice two *y*-axis with different scale) and zeta-potential of granular media (right) on the retention of viruses, phages and protozoan oocysts

Biofilm accumulation and natural organic matter (NOM) also affects the removal of biocolloids like oocysts as **Dai and Hozalski (2002)** demonstrated. They showed that NOM surface adsorption influences hydrophobicity of surfaces and electrostatic forces. From their results they concluded, however, that the change in hydrophobicity of the oocysts plays a minor role.

5.4.2.2 Contact time

Attachment efficiency in filter beds will be higher at higher contact times in the filter bed This was observed for bacteria and bacterial spores; higher DEC-values were observed in filter beds with higher contact times (Figures 5.3 and 5.5). For the other organisms this was not observed, partly because of a lack of information in some studies but also because it is not automatically so that a small sized filter bed with low filtration rate will have similar DEC value as a larger filter bed size with higher filtration rate. Higher filtration rates with higher shear forces may result in an increased breakthrough of micro-organisms.

Detachment of reversible attached micro-organisms caused by changed in water chemistry (pH and conductivity) may also attribute to the detachment behaviour as demonstrated in column studies. It will be difficult, however, to demonstrate this effect on breakthrough of attached micro-organisms and separate it from increased breakthrough caused by hydraulic effects in these dynamic filters as described hereafter.

5.4.3 Factors related to detachment: hydraulic conditions

The studies presented above were looking at the attachment behaviour. Detachment and breakthrough will be influenced by the hydraulic conditions like filtration rate, particle accumulation and stage of filter run (ripening, stable operation and breakthrough) and variations in water quality influencing the surface properties.

5.4.3.1 Filtration rate

The flow rate of the water in the granular media will affect the attachment of the organisms in the filter bed. For polioviruses Robeck *et al.* (1962) showed a decreasing removal at increasing filtration rates (Figure 5.15). The range of tested filtration rates in this study, however, was far below the rates applied at RGF-processes. Furthermore, because of the low effect of the filtration rate observed in the range between

0.1 and 8 m/h (shown in Figure 5.15), it is not expected that the filtration rate in the RGF range has a large impact on the removal of viruses by this process.

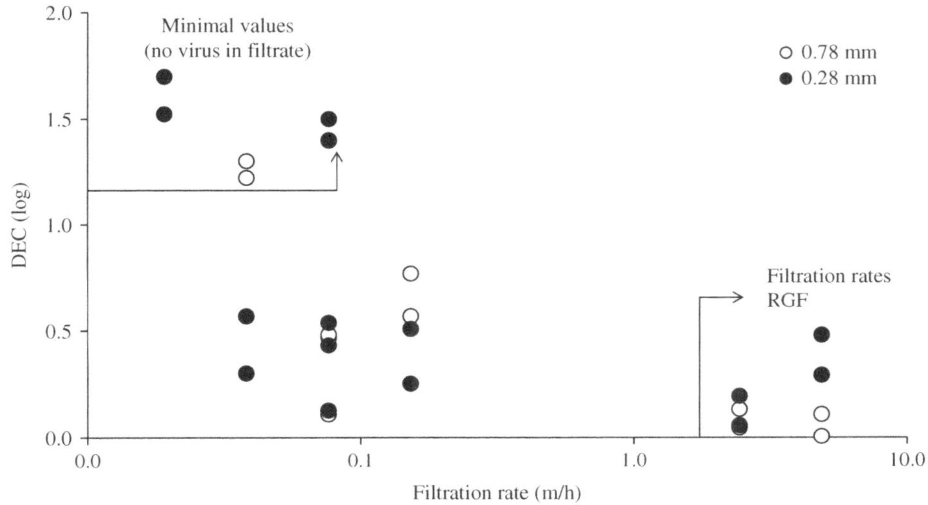

Figure 5.15 Relationship between removal of polioviruses by RGF and the filtration rate (Robeck *et al.*, 1962)

The effect of the filtration rate on the removal of *Cryptosporidium* was examined by relating the removal data of all literature studies used in this study with the applied filtration rates (Figure 5.16). No relation was observed for the RGF processes, without or with preceding coagulation.

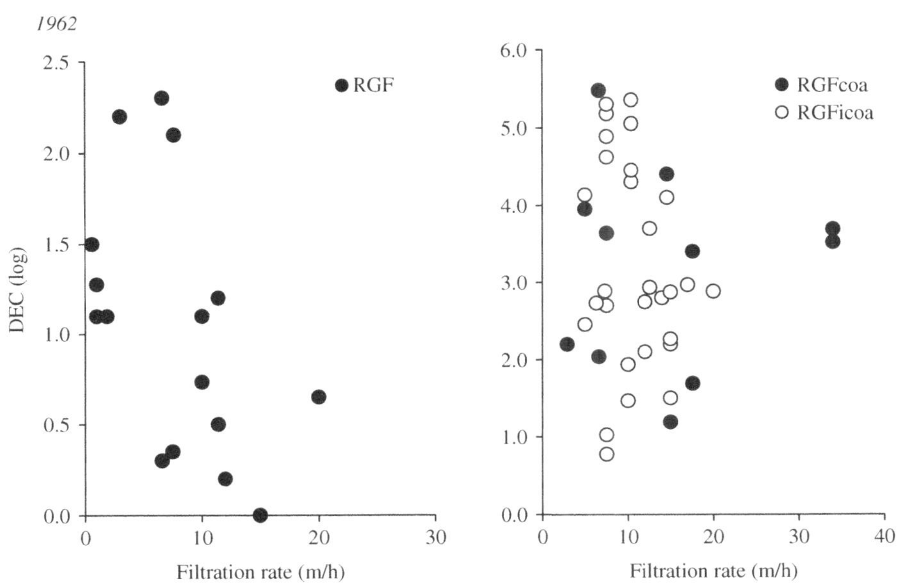

Figure 5.16 DEC-values of RGF processes for Cryptosporidium related to the filtration rate of the filters (all studies evaluated in this review)

Variations in filtration rate may occur during drinking water production and Emelko et *al.* 2001 showed in her study the effect of a hydraulic step of 25% on the removal of *Cryptosporidium* oocysts (Figure 5.17). In three separate filter runs it was observed that this effect can be moderate to very high ultimately leading to a negative removal rate with a DEC decrease from 5 to 3.5 log. This observation indicates that suddenly increasing flows in the filter bed can be sometimes responsible for a high percentage of detachment of oocysts accumulated in the filter bed. A similar effect was observed for the removal of bacterial spores of *B. subtilis* simultaneously dosed to the filter bed. Although the spores were removed to a lower extent than the oocysts, the detachment behaviour of these surrogates was predictive for the detachment behaviour of *Cryptosporidium*.

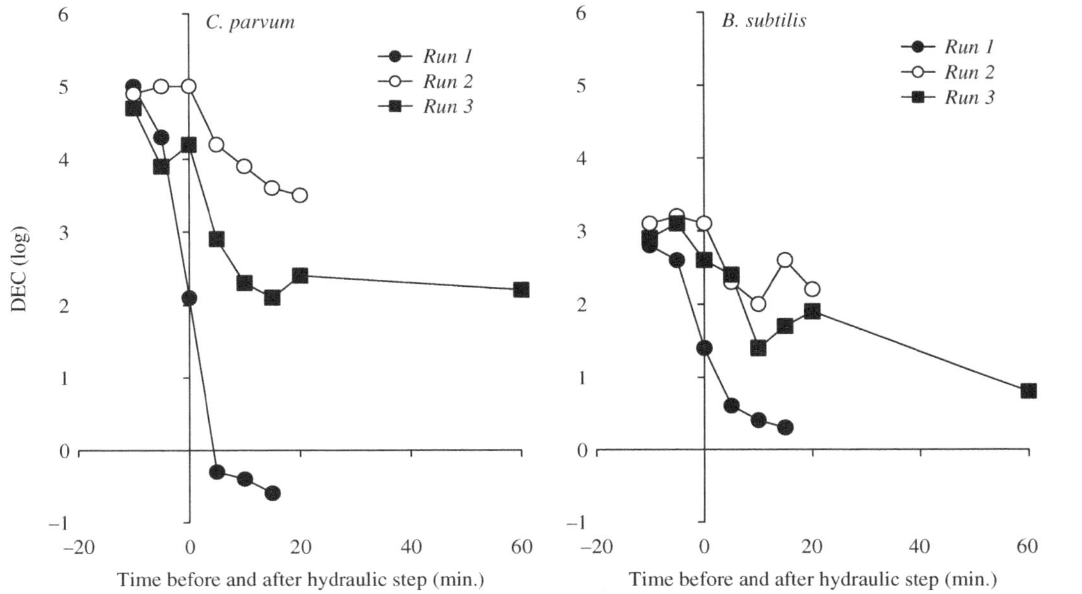

Figure 5.17 The removal of oocysts of *C. parvum* and spores of *B. subtilis* by a RGF (after coagulation/sedimentation) during the operational time before and after a sudden hydraulic step of 25% (from Emelko, 2001)

5.4.3.2 Filter media, preferential flow and back wash effect

As generally evaluated no systematic effect of the granular material in the filter beds on the DEC-values was observed. In two separate studies both a sand filter (Sa) and a dual media filter with anthracite/sand (AnSa) showed no difference in removal rate in both filters for either poliovirus or oocysts of *Cryptosporidium* (Figure 5.18). The studies of Hall *et al.* (1995), Swertfeger *et al.* (1999) and Emelko *et al.* (2001) for *Cryptosporidium* removal resulted in a similar conclusion.

Under the influence of different hydraulic conditions and a higher head loss development in the sand bed Robeck *et al.* (1962) demonstrated a quicker breakthrough of polioviruses compared to the dual media filter bed. Dugan *et al.* (2001) on the other hand showed that this effect can be diminished by the back wash strategy; by setting a limit to the maximum head loss in the filter bed, breakthrough of *C.parvum* in the sand filter was not observed (Figure 5.18). DEC of the sand filter and the dual medium filter was similar. These results show that filter beds vulnerable to head loss development, are more vulnerable for micro-organism breakthrough, but the effect can be minimised by optimizing back wash strategies.

The breakthrough behaviour of a filter bed is usually monitored by turbidity measurements as a parameter for breakthrough of suspended solids. Both Harrington *et al.* (2003) and Dugan *et al.* (2001) presented results of decreased *Cryptosporidium* removal by RGF during periods with increasing breakthrough of suspended solids monitored by turbidity measurements (Figure 5.19).

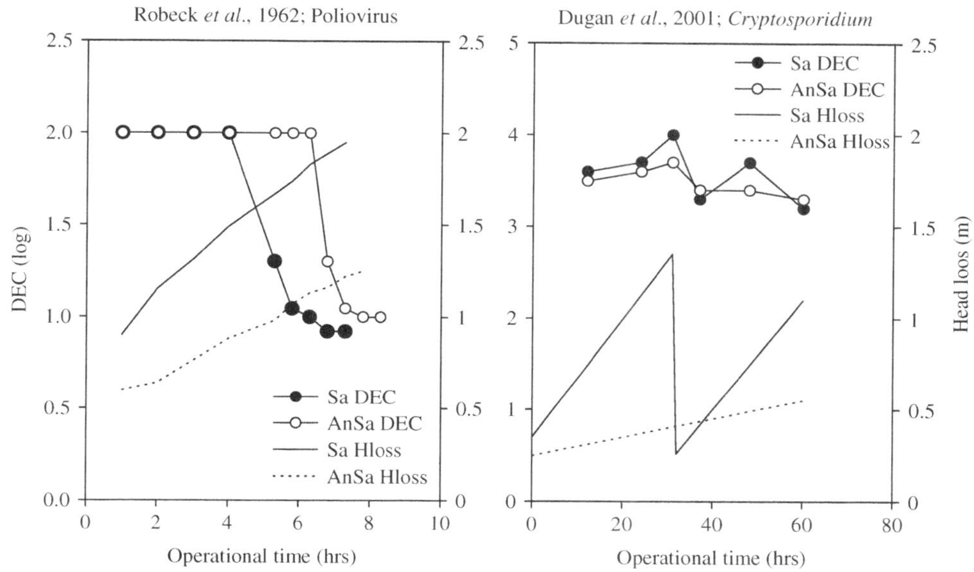

Figure 5.18 Removal of Poliovirus and Cryptosporidium by a sand (Sa) and dual-media filter (AnSa) in relation to the hydraulic conditions (head loss; Hloss) in the filter bed

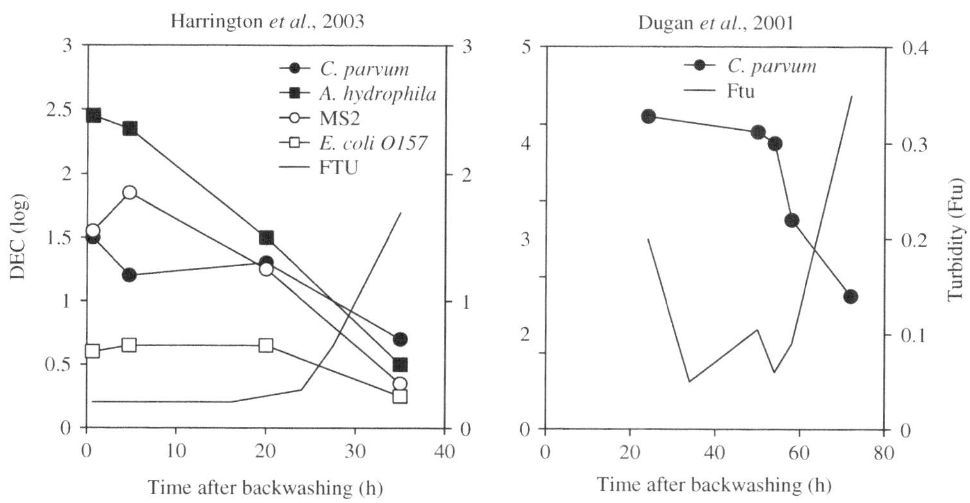

Figure 5.19 The relationship of micro-organism breakthrough and breakthrough of suspended solids measured by turbidity (notice different scales of left and right figures)

Emelko (2001) studied the oocyst, spore and particle removal systematically during different periods of the operational time of a RGFcoa, end of run, early breakthrough and late breakthrough (Figure 5.20). This clearly shows the impact of preferential flows in a filter bed on oocyst removal. Furthermore, the behaviour of the different particles is similar.

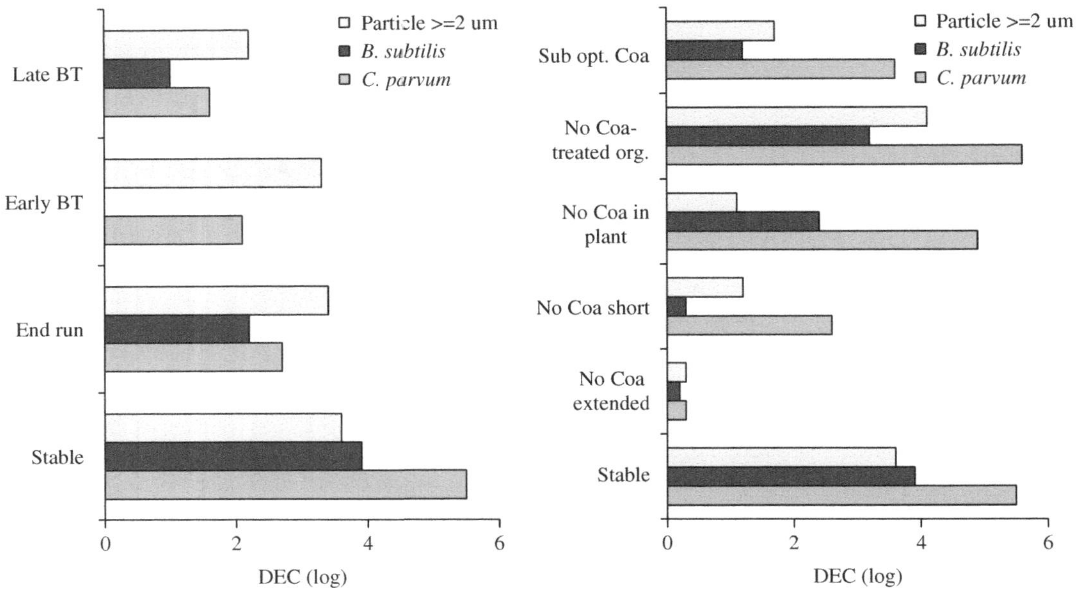

Figure 5.20 Influence of operational time (left) and (right) of the coagulation conditions on the removal of oocysts, particles and aerobic spores (from Emelko, 2001)

The impact of coagulation conditions, however, on oocyst, spore and particle removal was not similar (Figure 5.20). These changing conditions had a larger impact on oocysts removal than on removal of particles and spores. Furthermore, oocysts removal during a period when the oocysts were not coagulated before addition and during stable conditions were similar, whereas no coagulation in the treatment prior to the filter bed had a large impact on removal of particles, spores and also to some extent oocysts. This indicates that accumulated coagulant in the filter bed serves as a filter aid with increasing filtration efficiency.

All the studies were conducted with dosed micro-organisms. Two studies monitored removal of environmental bacteria (Coli44 and *E.coli*) and environmental bacterial spores (SSRC and aerobic spores) by RGF processes during operation. A considerable variation in DEC-values over time (Figure 5.21) was observed in both studies. In agreement with the overall MEC-values (Tables 5.2 and 5.3), data of Hijnen *et al.* (1997a) and Gray *et al.* (1999) showed generally a higher spore removal (SSRC and aerobic spores) compared to the removal of coliforms (Coli44 or *E. coli*). Furthermore, no quantitative relation was observed between the removal data of bacteria and bacterial spores. Only for the spores, Hijnen *et al.* (1997a) showed a short reduction in removal rate during the first few hours after backwashing. Most likely during the monitoring period of Gray *et al.* (1997) the filter beds with relative low filtration rates of 1.4 m/h were not backwashed (not described in the document).

5.4.4 Effects of coagulation conditions

In experiments Patania *et al.*, 1995 and Emelko (2001) showed that the conditions of a preceding coagulation have a significant effect on the MEC of RGF (Figure 5.22) for *Cryptosporidium*. Both authors presented extended studies on (oo)cyst removal. The average DEC of RGFcoa under conditions without coagulation (Patania) or failure in coagulation (Emelko) was significantly less than under normal coagulation conditions.

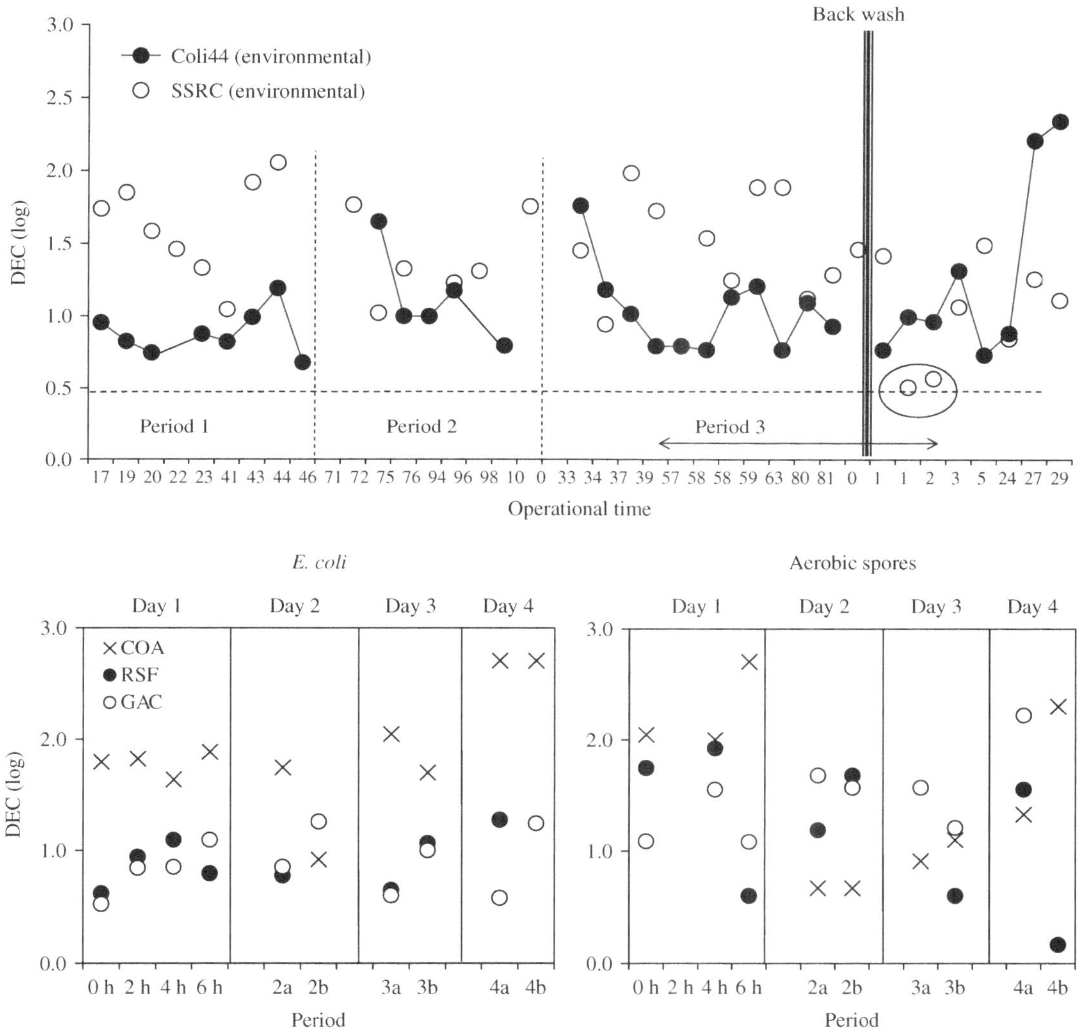

Figure 5.21 The DEC-value of RGF for Coli44 and SSRC (Hijnen *et al.*, 1997a) and E.coli and aerobic spores (Gray *et al.*, 1997) monitored during several periods of operational time

A large difference in DEC-values under stable conditions was observed between the studies of Patania and Emelko. This can be caused by the higher filtration rate and smaller sand bed in the Patania study compared to these conditions in the Emelko study. These differences in conditions (filtration rate and granular media), however, have been observed in the overall data set with no measurable effect on DEC. Therefore it is more likely that the difference in dosage strategy is the cause of the large difference: in the Patania study oocysts were dosed to the raw water before addition of the coagulants, whereas in the Emelko study coagulated oocysts suspensions were dosed in the influent of the filter bed. Furthermore, in a relative limited experiment Emelko demonstrated that pre-coagulation of the dosed oocysts under stable plant conditions did not affect DEC (see in Figure 5.20; no COA-treated org.). Several other authors also demonstrated the importance of proper coagulation conditions for the performance of RGF process to remove *Cryptosporidium* (i.e. Dugan *et al.*, 2001; Edzwald and Tobiason, 2002). Also the higher MEC-value of RGFcoa compared to RGF calculated for all groups of micro-organisms shows the enhanced effect of removal caused by coagulation.

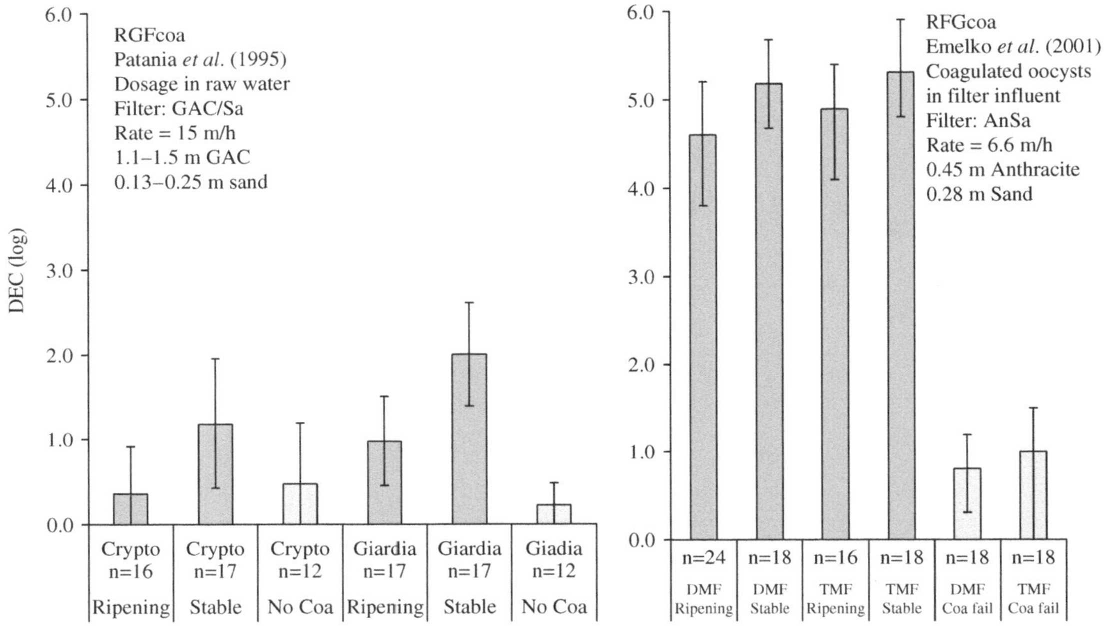

Figure 5.22 Effect of ripening and coagulation failure (RGFcoa) on the elimination of oocysts by RGFcoa (error bars = SD)

Most of presented studies are mainly focused on disturbances in coagulation conditions related to the floc formation caused by underdosing of chemicals. Also environmental conditions like temperature and water quality (i.e. turbidity, pH) influence floc formation and thus contribute to the variation in DEC of the subsequent RGF. Additionally, there are results presented which may be interpreted as results which contradict the conclusion on coagulation conditions. Hijnen *et al* (1997) showed that bacterial spore (SSRC) breakthrough observed after coagulation (Dissolved Air Flotation (DAF)) was compensated by a subsequent dual media filter (Figure 5.23). To some degree Gray *et al.* (1999) observed similar results for the removal of aerobic spores and *E.coli* by a sludge blanket clarifier (SBC) and sand filtration. These results were observed under normal and stable process operation, thus no obvious disturbances in coagulation conditions. It is hypothesised that this compensation by RGF only counts for micro-organism breakthrough after coagulation caused by variations in the primary floc removal and for breakthrough caused by failures in the floc formation conditions. All studies on the effect of failing coagulation conditions on DEC have been performed with special prepared and dosed oocysts. The above presented studies were based on removal data of environmental spores and *E. coli*. Differences in conditions of the removed micro-organisms may affect fate of these micro-organisms during the coagulation/filtration process. To verify this, additional data on the removal of environmental oocysts by the separate processes in conventional treatment are needed.

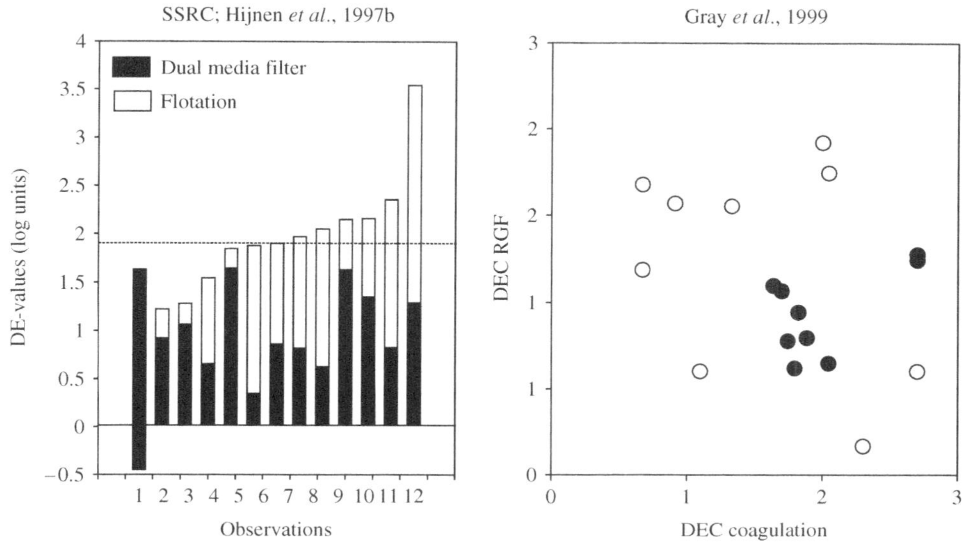

Figure 5.23 Synergetic effect of coagulation and secondary floc removal in RGF; low removal in coagulation is compensated by higher removal in RGF

5.5 Surrogates for pathogen removal by RGF

In literature several environmental parameters have been suggested as surrogates for the removal of pathogenic viruses, bacteria and protozoan (oo)cysts by filtration processes. For viruses bacteriophages like F-specific RNA-phages or somatic coliphages have been investigated, for pathogenic bacteria like *Campylobacter* indicator bacteria like *E. coli* and enterococci are potential surrogates and for *Giardia* and *Cryptosporidium* particles, algae and (an)aerobic spores have been proposed. From the collected data sets, the MEC-values of these separate groups of organisms for the different types of RGF processes are calculated (Figure 5.24).

From these data it is obvious that phages are proper surrogates for the removal of viruses by the three different filtration types. From the limited data on algae removal it can be concluded that the DEC of RGF-processes assessed by these environmental organisms comes close to the DEC-values assessed for protozoan (oo)cysts (with dosage experiments). Except for RGF, the values calculated for bacterial spores were similar to the results of the algae. Aerobic spores in combination with particle count, microspheres and turbidity as surrogates have been studies intensively by several authors in dosing studies (Coffey *et al.*, 1999; Dugan *et al.*, 2001; Emelko *et al.*, 2001; Patania *et al.*, 1995; Swertfeger *et al.*, 1999; Yates *et al.*, 1997a). General trend in these studies is that spores, particles and turbidity removal underestimates the removal of (oo)cysts by RGF. This was demonstrated for particle count and aerobic spores for instance by Emelko *et al.* (2001; Figure 5.25). Additionally, the latter study demonstrated that removal of specific dosed microspheres was proportional to the removal of oocysts of *C. parvum* (Figure 5.25). Furthermore, turbidity and particles are found to be more conservative than the spores. Breakthrough behaviour monitored with these surrogates reflects mostly the breakthrough behaviour of the (oo)cysts. Thus, these parameters can be used during process optimisation. An example for this is shown in Figure 5.17. Also the anaerobic spores (SSRC or *C. perfringens*) are found to be conservative surrogates for protozoan (oo)cyst removal by filtration as shown in other filtration studies (soil column study (Hijnen *et al.*, 2005) and slow sand studies (Hijnen *et al.*, 2004, 2006)).

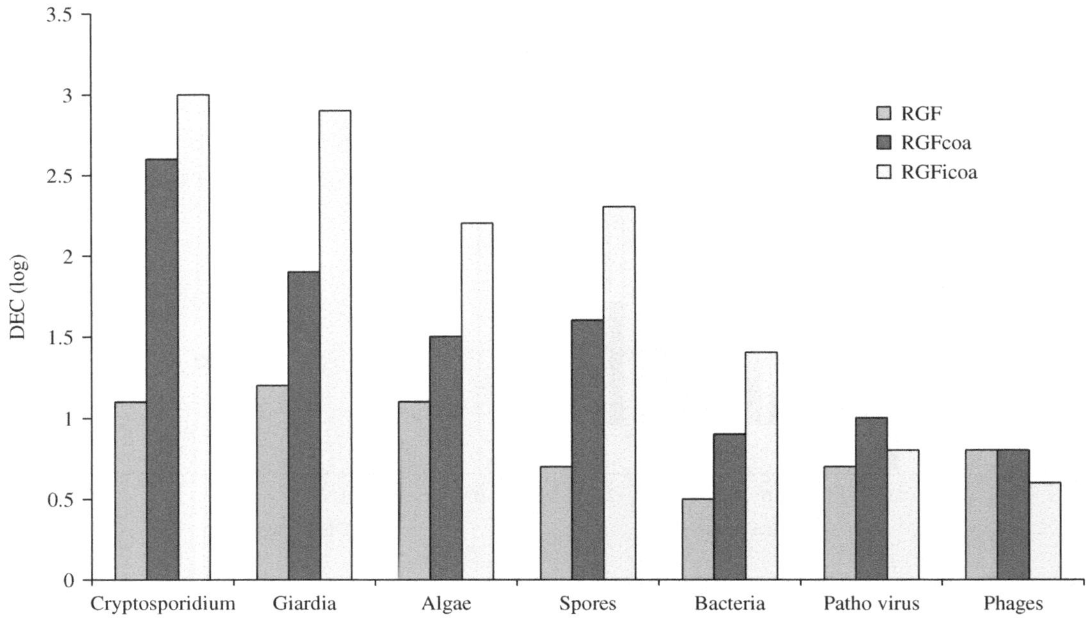

Figure 5.24 Comparison of the MEC-values of RGF processes for the different kind of micro-organisms

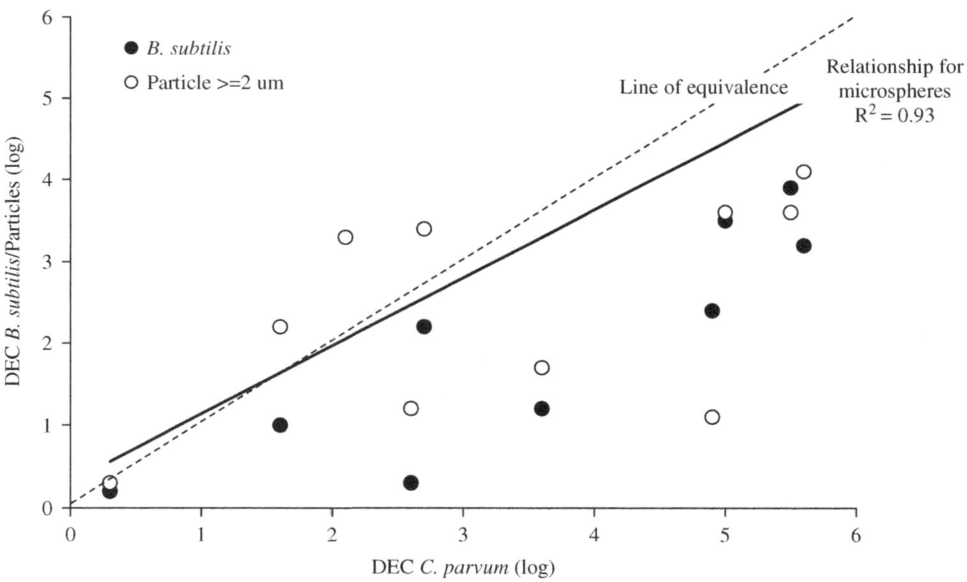

Figure 5.25 Relation between *C. parvum* removal and removal of spores of *B. subtilis* and particles (> = 2 um) (Emelko (2001)

The use of particle counters, off-line and in-line for treatment assessment was studied in the Netherlands by Hijnen *et al.* (1999). It was concluded that off-line instruments could be used easily at different points in the treatment to assess DEC-values for particles, whereas in-line instruments are more useful for the registration of dynamics in processes as presented by others. There was a significant correlation ($P < 0.001$) between particle count (off-line instruments), although not proportional, and general biomass parameters like ATP, direct microscopic cell count (TDC) and turbidity. Furthermore, the predicted DEC-values of conventional treatment (coagulation/filtration) and RGFgac for particles were more similar to the DEC values for SSRC than to the DEC values for turbidity and both biomass parameters (Figure 5.26). Negative DEC or particle release as described in paragraph 5.2 for spore removal by RGFgac was detected by all parameters but turbidity.

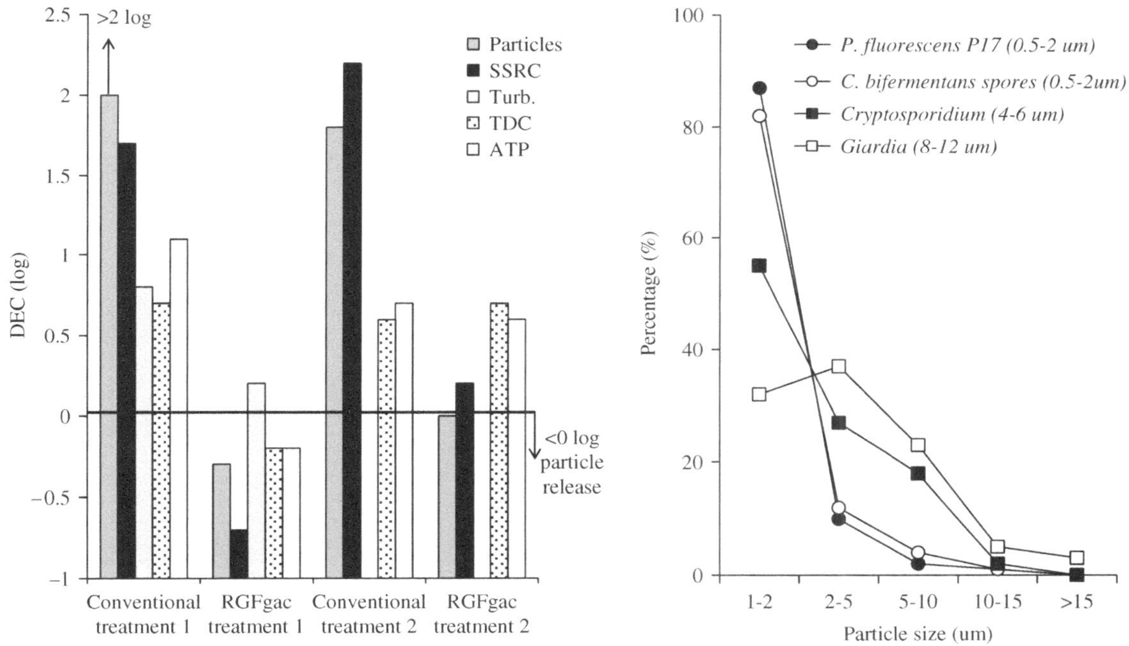

Figure 5.26 Particle removal (off line with light blockage sensor) compared to biomass removal (ATP, TDC), turbidity and SSRC (left) and (right) particle size distribution observed with a light blockage sensor in particle poor water dosed with bacteria, bacterial spores and protozoan (oo)cysts (from Hijnen *et al.*, 1999)

With the off-line monitor it was demonstrated that light blockage sensors underestimate the size of all relevant microbial cells (Figure 5.26), a conclusion also presented by O'Shaughnessy *et al.* (1997). Though on-line monitors can also give valuable information on process performances, operational problems were encountered with the continuous flow and fouling of the sensor and/or tubing.

For slow sand filtration *E. coli* is a conservative surrogate for removal of *Campylobacter* as demonstrated by Hijnen *et al.* (2004) and Dullemont *et al.* (2006). For rapid sand filtration, however, the study of Hijnen *et al.* (1998d) showed that *E. coli* is removed slightly better than environmental *Campylobacter* bacteria.

5.6 Conclusions and further research

MEC of RGF and comparisson with conventional treatment

From the MEC-values of RGF-processes for the different groups of micro-organisms it can be concluded that the process contributes to the overall removal of micro-organisms in water treatment. This contribution varies and depends on the type of micro-organism and on the additional use of a preceding coagulation. Without a preceding coagulation the MEC range from 0.5 log for the bacteria to 1.2 log for *Giardia* cysts (Figure 5.13). RGF is commonly used as a secondary floc-removal in conventional treatment and under theses conditions MEC for bacteria and viruses is slightly (0.4 log) higher than RGF without a preceding coagulation (Figure 5.27).

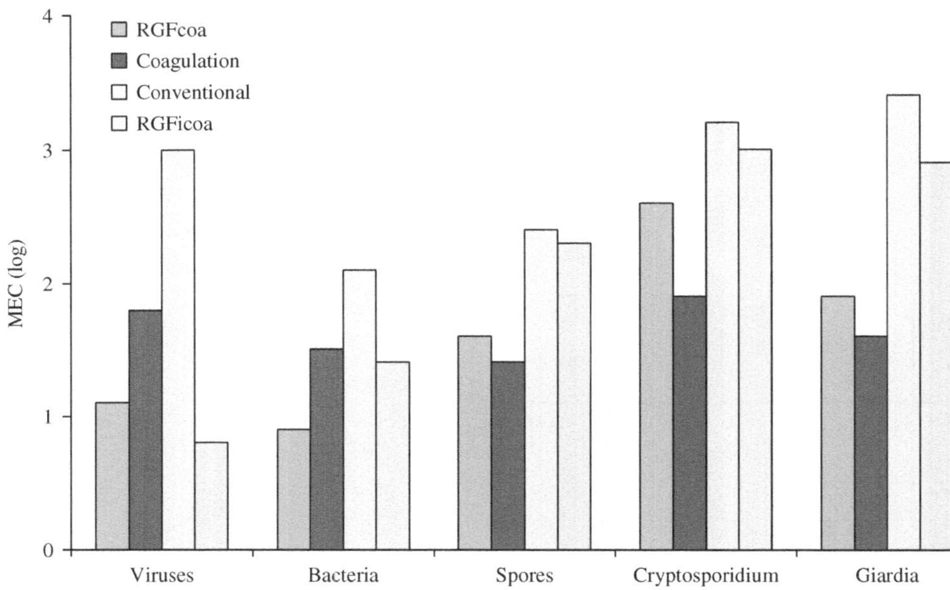

Figure 5.27 Comparison of MEC-values for the individual processes coagulation and RGFcoa with the MEC-values calculated for the combination of the processes separately applied (conventional treatment) and or integrated (RGFicoa)

For bacterial spores, *Cryptosporidium* and *Giardia* higher the impact of a preceding coagulation process is higher (0.7–1.4 log). The enhanced efficacy of RGF by coagulation for spore and (oo)cyst removal compared to viruses and bacteria was also observed for RGFicoa, the combination of coagulation and filtration by adding coagulants to the influent of the filter bed. Especially for viruses this was not very effective but also MEC of RGFicoa for bacteria was lower than MEC of conventional treatment.

MEC values of RGF with granular activated carbon (RGFgac) for the different groups of micro-organisms was in the same order as observed for RGF without preceding coagulation process.

Influence of process conditions on MEC of RGF

In the data set the major process conditions were recorded if available and correlated qualitatively with the variation in observed DEC-values (Table 5.6). This revealed the following general conclusions:

- the effect of coagulation on the MEC of RGF is deduced from the differences in MEC-values between the different evaluated processes (Figures 5.13 and 5.27). A preceeding is positive for the elimination of viruses and bacteria but the effect was higher for the protozoan (oo)cysts. This effect was enlarged for bacteria and protozoan (oo)cysts by in line coagulation but not for viruses. For the latter organisms direct filtration of in line coagulation seems to be not an effective process;

- attachment is the major process mechanism for the removal of viruses in RGF whereas bacteria and protozoan (oocysts) will also be removed by straining. Physical/chemical conditions of the water and surface properties of the viruses and the granular media affect the elimination. Controlable conditions like pH of the water and the zeta-potential of the granular media will affect the elimination of viruses significantly; high pH cause low removal and high (positive) zeta-potential material high removal. For bacteria and protozoan (oo)cysts no specific studies were found showing the effect of these conditions on the DEC of RGF;
- as to the effect of granular media on the DEC-values of RGF in most studies no significant effect of the different traditional used media (sand/anthracite) was observed for *Cryptosporidium* and also virus removal. When a negative effect was observed this was attributed to differences in hydraulic conditions; the observed DEC-reduction could be avoided by optimization of the backwash strategy. A slightly lower MEC-value of granular activated carbon filtration was calculated compared to the MEC of RGF with the traditional media but the difference was not statistically significant;
- the effect of filtration rate or contact time was evaluated in the current review and showed that in the range of the common applied rates no systematic effect on the elimination of viruses and protozoan (oo)cysts was observed;
- hydraulic conditions of the filter bed do have a significant effect on the breakthrough behavior as shown for all micro-organisms and will reduce the average DEC-value considerably. The hydraulic conditions of the filter bed are related to the length of the ripening period after back washing, preferential flow caused by head loss or other operational conditions like hydraulic steps (sudden filtration rate variations).

Table 5.6 The influence of process conditions on the elimination of the different pathogens: available knowledge in the current review (+ yes and – no) and correlation (positive or negative correlation between parameter and DEC; + or -) plus the extent of the correlation (++ or – –)

	Viruses	*Bacteria*	*Cryptosporidium/Giardia*
Coagulation			
– preceding (RGFcoa)	+/+	+/+	+/++
– integrated (RGFicoa)	+/–	+/+	+/++
Attachment conditions			
pH	+/–	–	–
zeta potentail media	+/+	–	–
Granular media	+/–	–	+/–
Contact time/filtration rate	+/–	+/–	+/–
Hydraulic state of the filter bed[a] (preferential flow)	++	–	+/–

[a]Related to backwash strategy and filtration rate steps

Model organisms and surrogates to study behaviour of pathogenic micro-organisms in RGF processes are phages, indicator bacteria like *E. coli* and enterococci, anaerobic and aerobic bacterial spores, turbidity and particles. The latter two physical parameters and spores are useful parameters to optimise RGF processes for the removal of *Cryptosporidium* and *Giardia*, the two most emerging pathogens in water treatment for which filtration is a major barrier in water treatment.

Further research

- Extended RGF research on microbial removal has been executed in the last decade as demonstrated in this chapter. These data indicate that the major process optimisation aspects are known and are connected to operational conditions like stable operation of the filtration rate and back wash strategies. This knowledge can be implemented on local treatment plants to optimise the process performance. These optimisations must be focused on reducing process variability of turbidity and particle removal, thus increasing average DEC (MEC) and reducing process failures which may result in increasing infection risks when occurring simultaneously with peaks in pathogen concentrations in the source water;
- Because many of the presented studies used dosage of pre-cultured organisms, one of the major challenges in further research is to collect more data on the removal of environmental viruses, bacteria and protozoan (oo)cysts to validate the calculated MEC-values in this report.

6 Slow sand filtration

Slow sand filtration is one of the oldest water treatment processes used to produce microbiologically safe drinking water. Application of this process was effective against waterborne outbreak of cholera in Hamburg as demonstrated by Koch in the 19th century (Koch, 1893). Nevertheless, quantitative data about the elimination capacity of this process in literature on viruses and protozoan oocysts are limited and show a high degree of variation (Hijnen and Schijven, 2001). Additional research was conducted by Amsterdam Water Supply and Dune Waterworks of South-Holland in cooperation with RIVM and Kiwa Water Research (Dullemont *et al.*, 2004, 2006; Hijnen *et al.*, 2004, 2006a, 2007; Schijven *et al.*, 2008; Visser *et al.*, 2004) to collect more quantitative information.

6.1 Viruses

For virus removal by slow sand filtration (SSF) nine studies were found. Only one study (Slade, 1978) was performed under full-scale conditions with environmental enteroviruses (FS = 5). The other studies were carried out with bacteriophages on pilot plant-scale or laboratory-scale. Therefore the overall FS-index was low (3.3; Table 6.1).

The oldest studies are those of Poynter and Slade (1977), Robeck *et al.* (1962), Slade (1978) and Windle-Taylor (1969). Due to little process information the studies of Wheeler *et al.* (1988) and Yahya *et al.* (1993) were not included. In the study of Robeck *et al.* (1962) poliovirus removal was studied under different filtration conditions; the conditions and results were not well described and removal ranged between 0.1 and 1.5 log. The Windle-Taylor study was not well documented with polioviruses added to the influent of a full-scale filter. The average removal was 3 log. Poynter and Slade presented a pilot plant study where the average removal of spiked polioviruses and bacteriophages was 2.8 (standard deviation (SD) of 0.7) and 3.5 (SD = 0.6) log, respectively. Slade monitored concentrations of environmental viruses in filtered water of full-scale filters. At temperatures of 6–7°C viruses were detected in 7 of the total of 19 samples. Based on the average concentrations before and after the filters a removal of 1.9 log was calculated.

McConnell *et al.* (1984) described filtration tests with reoviruses with and without radioactive-labelling. They showed an effective removal of viable reoviruses of 4.0 log in a 1.2 m filter bed with sieved and unsieved sand (effective sizes 0.28 and 0.18 mm and Uc 2.1 and 4.4, respectively). Labelled reoviruses, however, showed a DEC of 1.3 log (1.2–1.5 log; n = 4) when the numbers were determined with gamma-counting. Denny and Broberg (1992) studied the removal of indicator bacteria and coliphages by full-scale treatment and used large volume sampling technique. They determined a DEC of 2.1 log for coliphages. Hijnen *et al.* (2004), Wubbels (2003) and Dullemont *et al.* (2006) studied the removal of MS2 phages, a conservative model organism for viruses (Schijven, 2001), by pilot plant filters. DEC-values of 2.1, 0.9, 0.6 and 3.4 log were observed.

Based on these studies the MEC of SSF for viruses is 2.2 log. In Figure 6.1 all DEC-values included in the calculation are arranged in declining order and related to the hydraulic retention time (HRT; hrs.) in the filters as a measure of flow rate or filter bed depth. Schijven (2001) presented a comprehensive review on virus removal by soil passage where he showed theoretically and experimentally that removal increases with increasing residence time. DEC of SSF for viruses ranged from 0.9 up to 4.0 log but no relationship with HRT was observed. This can be caused by variation in type and state of organisms, in temperature and in quality of water and filter sand. 4.0 log removal of pre-cultured reoviruses was attributed to inactivation (McConnell *et al.* (1984). They observed a large difference between the removal of radioactive-labelled reoviruses monitored with a gamma-counter (1.2 log) and non-radioactive-labelled reoviruses monitored with immunofluorescent cell count (4 log). High DEC-values of Poynter and Slade (1977) and Windle-Taylor (1969) were monitored in open filters with a HRT of 1.5 and 1.7–9 hours. Inactivation of fluenced organisms due to natural UV-irradiation, biological activity (algae) in the water above the filter bed and high accumulation of organic matter on the filter bed surface are possible explanations for the high DEC of these filters. The studies by Dullemont *et al.* (2006) and Schijven *et al.* (2007) with covered SSF loaded with low-nutrient feed water demonstrated that the schmutzdecke did not affect the DEC for MS2-phages under these conditions. Low DEC-values were obtained from studies with MS2 phages in a pilot plant (Wubbels, 2000).

Table 6.1 MEC of slow sand filtration for micro-organisms

Organisms	Data characteristics			MEC (log)		
	Studies	Data	FS-index[s]	Average	P50	Range
Viruses	10[a,c,d,e,f,h,l,p,q,s]	13	3.3	2.2 (±1.1)	2.1	0.6–4.0
Bacteria	9[d,e,g,i,j,l,r,s]	17	3.4	2.7 (±1.1)	2.4	1.2–4.8
Bacterial spores	5[b,l,m,q,t]	9	4.4	1.5 (±1.3)	1.3	0.0–3.6
Cryptosporidium	5[k,n,o,q,t]	6	3.3	4.8 (±1.3)	nd	2.7–6.5
Giardia	3[i,k,t]	3	3.3	4.9 (±1.0)	nd	4.0–6.0

[a]Robeck *et al.*, 1962; [b]Burman *et al.*, 1962; [c]Windle-Taylor *et al.*, 1969; [d]Poynter *et al.*, 1977; [e]Slade *et al.*, 1978; [f]McConnell *et al.*, 1984; [g]Cleasby *et al.*, 1984; [h]Ellis, 1985; [i]Bellamy *et al.*, 1985; [j]Schuler *et al.*, 1991;[k]Denny *et al.*, 1992; [l]WRC, 1995; [m]Timms *et al.*, 1995; [n]Wubbels, 2000; [o]Hijnen *et al.*, 2003; [p]Hijnen *et al.*, 2004; [q]Dullemont *et al.*, 2006; [r]Hijnen *et al.*, 2006; [s]the higher the number, the more equivalent with full-scale situation

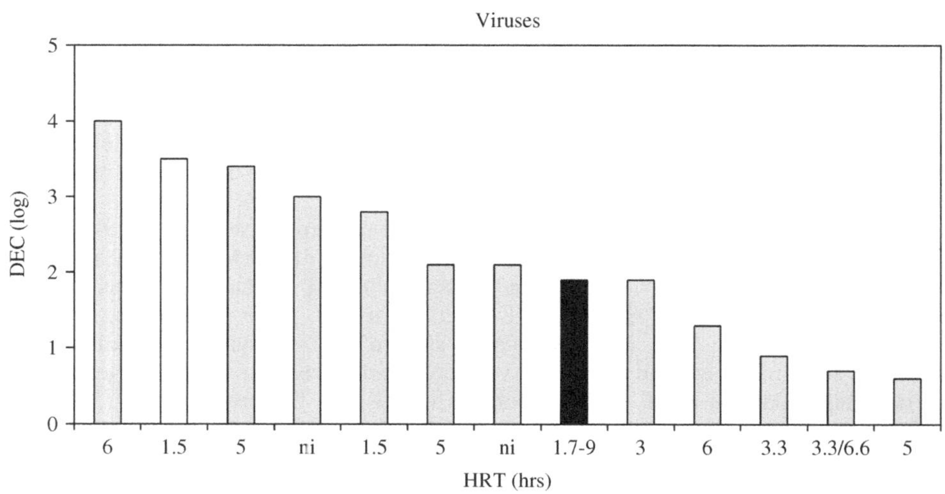

Figure 6.1 DEC of slow sand filtration (SSF) for viruses and bacteriophages arranged in declining order (Black, grey and white bars: FS-index of 5, 4/3 and 2/1 respectively)

6.2 Bacteria and bacterial spores

Data on the removal of *E.coli* bacteria by SSF came from eight studies (Bellamy *et al.*, 1985; Cleasby *et al.*, 1984; Ellis, 1985; Poynter and Slade, 1977; Slade, 1978). Removal of thermotolerant coliforms (Coli44) and faecal streptococci in full-scale treatment was monitored by Denny and Broberg (1992), Dullemont *et al.* (2006), Hijnen *et al.* (2004) and Schijven *et al.* (2008). The latter determined also the elimination of *E.coli* and *Campylobacter* bacteria under pilot plant and laboratory conditions. The overall FS-value for bacteria was relatively low (3.4; Table 6.1) because most data came from full-scale studies where the elimination of non-pathogenic bacteria as surrogates was monitored.

A MEC of 2.7 log was calculated for removal of pathogenic bacteria by SSF and similar to the viruses no relation was observed with the HRT (Figure 6.2). The highest DEC-value of 4.8 log was observed for the removal of *Campylobacter lari* and *E. coli* bacteria spiked to a pilot plant filter. Environmental *Campylobacter* bacteria were removed with 3.4 log in full-scale filters on an average (Hijnen *et al.*, 2004) and the same filters removed Coli44 with 2.5 log, which suggests that these indicator bacteria can be used as a conservative surrogate for the removal of these pathogenic bacteria. Dullemont *et al.* (2006) came to the same conclusion based on spiking experiments with *E. coli* and *Campylobacter lari*. The DEC for bacteria ranged from 1.2 up to 4.8 log. Denny and Broberg (1992) observed a low removal for faecal streptococci by full-scale filters. Poynter and Slade (1977) showed that elimination was low under the following conditions: high filtration rate and low hydraulic retention time, filter beds with fresh sand and at low temperatures. Variation in the efficacy of individual filters for *Campylobacter* and Coli44 removal was presented by Visser *et al.* (2004). They attributed this variation to preferential flow caused by disintegration of joints in the concrete construction and/or digging and refilling of holes for sampling of the filter bed during maintenance. The same study also showed the importance of the schmutzdecke and of the application of a filter-to-waste period (FTW) after surface cleaning; the DEC calculated from all samples (including the FTW period) was for most individual filters lower than the DEC calculated from the samples taken during the production period only (Table 6.2). The pronounced effect of the schmutzdecke was also demonstrated for the removal of spiked *E. coli* on pilot plant scale (Hijnen *et al.*, 2004 and Dullemont *et al.*, 2006).

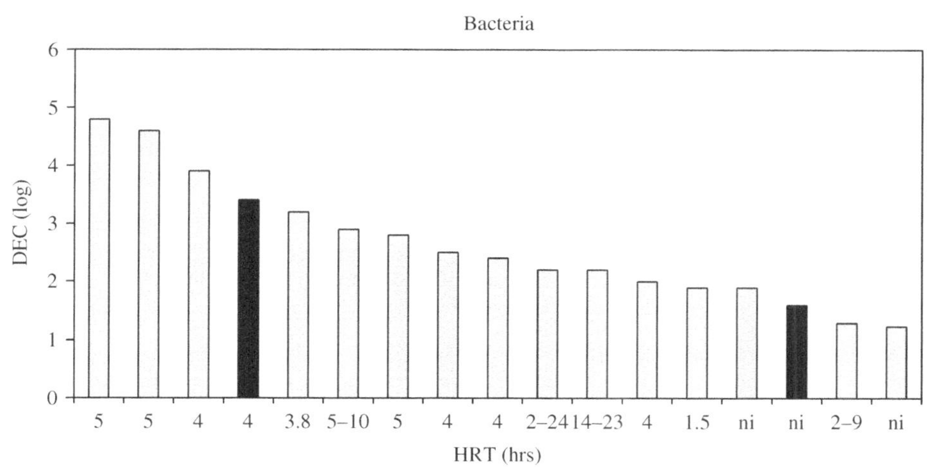

Figure 6.2 DEC of SSF for bacteria arranged in declining order (*Black, grey and white bars: FS-index of 5, 4/3 and 2/1 respectively*)

Table 6.2 DEC of individual filters calculated from samples collected in a period with or without filter-to-waste FTW (*from Visser et al., 2004*)

Individual filters	Campylobacter		Coli44	
	With FTW period[a]	Without FTW period	With FTW period	Without FTW period
1	>3.5	>3.5	1.9	2.3
2	3	3	2.3	2.3
3	2.8	3.2	1.5	1.6
4	2.1	>3.6	1	2.4
6	1.6	2.7	1.4	2
8	3.1	3.1	1.5	2.9

[a]Filter-to-Waste period: period with breakthrough of micro-organisms

Only few studies presented the removal of bacterial spores by SSF. Most of data used in this review came from Hijnen *et al.* (2004; 2006; 2007) who described the removal of environmental spores of sulphite-reducing clostridia (SSRC) by full-scale filters, which ranged from 0.0 up to 1.5 log (Figure 6.3). Higher removal of SSRC by SSF was observed by Water Research Centre (WRC) (1995). Much higher was the elimination of lab-cultivated spores of *C. perfringens* by columns filled with sand from full-scale filters (2.8 log; Hijnen *et al.*, 2004) and by a pilot plant filter (4 log; Hijnen *et al.*, 2006a).

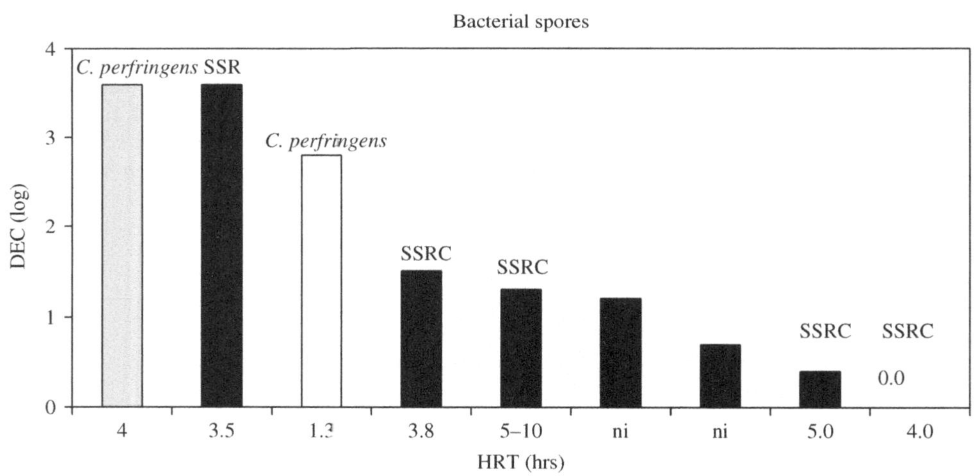

Figure 6.3 DEC of SSF for bacterial spores (spores of sulphite-reducing clostridia SSRC) arranged in declining order (Black, grey and white bars: FS-index of 5, 4/3 and 2/1 respectively)

The low and sometimes negative DEC observed for SSRC in full-scale filters is, compared to the high removal of spiked spores, possibly indicative for the removal of other persistent micro-organisms like *Cryptosporidium*. Multiplication of anaerobic spores as demonstrated by Willis (1956) in open filters in England (no cover above the sand bed) is not plausible for these Dutch SSF. Anaerobic conditions necessary for multiplication is prevented in these filters by a roof above the filters (no algae bloom) and by avoiding production stops. Furthermore, the influent is generally intensively pretreated surface water and therefore low in nutrients. Nevertheless, regrowth can not fully be excluded as a possible explanation.

A more plausible hypothesis for the low DEC for environmental spores is that spores reversibly attach to the sand, slowly move in the filter bed after detachment and persist long enough to pass the sand filter in the long run. This mechanism is called "accumulation and retarded transport (delayed transport)". This means that the high DEC-values for these persistent organisms calculated from short challenge tests with high concentrations are not applicable for full-scale filters in general. DEC will decrease over time and the health significance of this observation is depended on the rate of decrease of DEC and the rate of survival of the persistent pathogens.

6.3 *Cryptosporidium* and *Giardia*

Seven studies were found which described removal of protozoan (oo)cysts (*Cryptosporidium* and *Giardia*) by slow sand filters. Six of these studies investigated removal of *Cryptosporidium* (Fogel *et al.*, 1993; Hijnen *et al.*, 2004, 2006a, 2007; Schuler *et al.*, 1991; Timms *et al.*, 1995; WRC, 1995). Some authors (Schuler and Fogel) and also Bellamy *et al.* (1985) presented data on *Giardia* cysts removal. Only Fogel *et al.* (1993) monitored removal of environmental oocysts at full-scale conditions (FS = 5). The average FS-index was in same order of magnitude as FS-values calculated for the viruses and bacteria studies (Table 6.1).

Five of the six *Cryptosporidium* studies were challenge studies and in three of those no breakthrough was observed (Figure 6.4). In all presented *Giardia* studies cysts were observed in the filtrate and the DEC-values could be calculated (Figure 6.5).

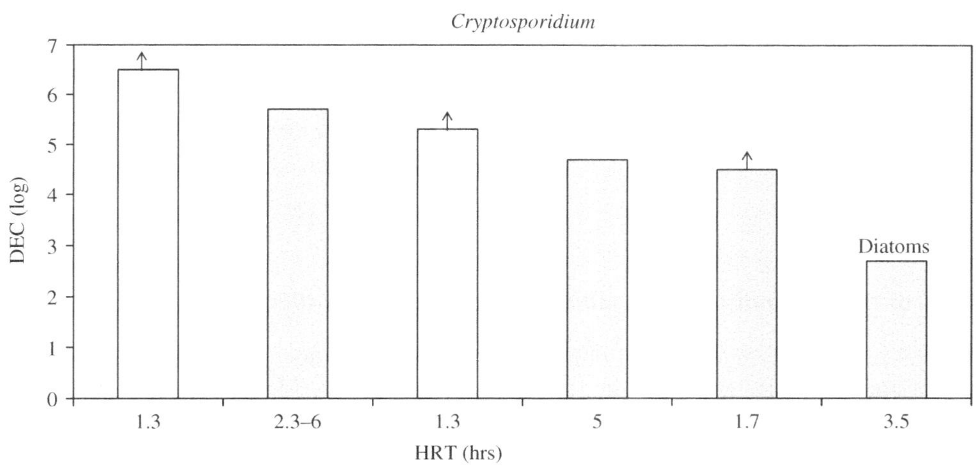

Figure 6.4 DEC of SSF for oocysts of Cryptosporidium arranged in declining order (Arrow means minimal DEC-value: no oocysts in filtrate; Black, grey and white bars: FS-index of 5, 4/3 and 2/1 respectively)

Most studies yielded high DEC-values. The studies with low DEC-values investigated environmental *Cryptosporidium* and *Giardia* (oo)cysts removal (0.3 and 1.2 log, respectively; Fogel *et al.*, 1993) and removal of centric diatoms (Hijnen *et al.*, 2006, 2007) by pilot plant filters (1.8 log). Combining these results with similar observations for the persistent clostridial spores leads to the hypothesis that peak-concentrations of protozoan (oo)cysts will be removed very efficiently and instantaneously by well designed and relatively young full-scale filters without preferential flows. However, due to the same mechanism of accumulation and retarded transport as described for clostridial spores before, DEC of full-scale filters may fluctuate and will reduce over time depending on the load, the rate of transport and rate of inactivation as described before. In the extended study after the fate of these persistent micro-organisms in the filter bed (Hijnen *et al.*, 2006a, 2007), however, it was concluded that the change of an increase of DEC over time due to delayed transport of oocysts is most likely low due to predation. The low DEC observed by Fogel *et al.* (1993) was based on breakthrough of oocysts during winter time at low temperature. Therefore DEC calculated on the basis of the average concentrations of influent and effluent data of the Fogel study is less valid and as such not included in the MEC-calculation. Based on these values a MEC of 4.8 was calculated for *Cryptosporidium* (Table 6.1). Assuming predation affects *Giardia* removal in the same way as *Cryptosporidium* removal and the DEC observed for *Cryptosporidium* by Hijnen *et al.* (2006a, 2007) is also applicable for the larger *Giardia* cysts a MEC-value of 4.9 log was calculated (excluding the low DEC of Fogel *et al.* (1993) from the data).

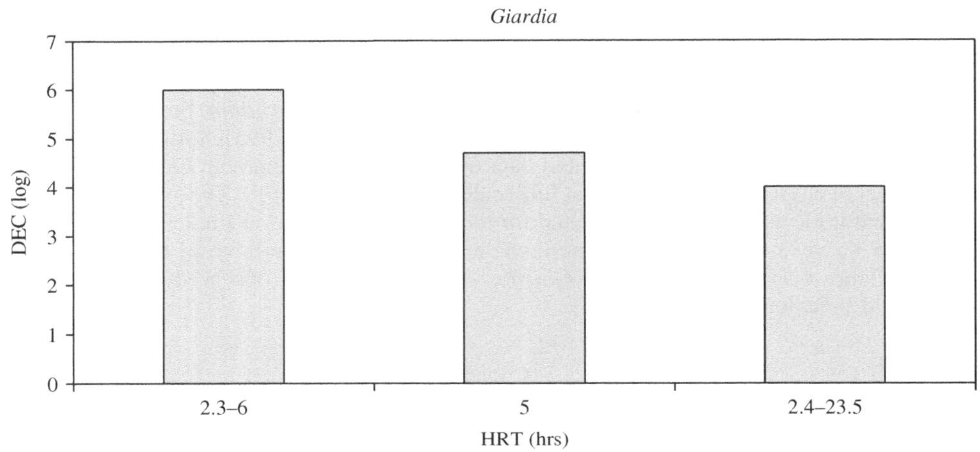

Figure 6.5 DEC of SSF for cysts of Giardia arranged in declining order (Black, grey and white bars: FS-index of 5, 4/3 and 2/1 respectively)

6.4 Discussion

6.4.1 MEC of the different micro-organisms and process conditions

This review revealed an average removal of bacteria and viruses by slow sand filtration of 2–3 log, which was in the same range as estimated in the literature review of Medema and Theunissen (1996). The schmutzdecke, water quality and grain size did not influence the removal of MS2 phages (Dullemont *et al.*, 2006; Hijnen *et al.*, 2004). Adsorption and decay are probably the main mechanisms, which govern removal of viruses in filter beds, which implicates that removal will be influenced by temperature, retention time and physical/chemical characteristics of sand and water.

Bacteria will be removed for some part by adsorption and decay but also straining is a probable removal mechanism (Schijven *et al.*, 2007). Therefore besides temperature, retention time and physical/chemical conditions of sand and water also schmutzdecke and grain size will affect the efficacy of the process. The effect of both the schmutzdecke and temperature on DEC of pilot plant filters operated under similar conditions is presented in Figure 6.6. Remarkable was the high impact of temperature on DEC for MS2 phages and *E. coli* in a relative small temperature range (7–15°C). In a period of approximately 50 days or less the high DEC for *E. coli* before filter bed scraping was restored.

Medema and Theunissen (1996) estimated a wide range (1.2–>3.7 log) of elimination of protozoan oocysts of *Cryptosporidium* and *Giardia* by slow sand filters. This was confirmed by the results presented in this chapter. As shown in Table 6.1 protozoan oocysts will be eliminated more effectively by SSF than bacteria or bacterial spores and more than for the other organisms straining will be an important mechanism (Bradford *et al.*, 2003, 2004 and Tufenkji *et al.*, 2004; Schijven *et al.*, 2007). These organisms will be trapped in small pores of the sand column which will implicate that removal is irreversible and oocysts that pass the sand column, travel through "the highways" formed by large sized pores. Hydraulic characteristics of the material like average grain size and grain size distribution, porosity and percentage of fines affect the removal efficiency. Logan *et al.* (2001) came to the same conclusion from their column study with both fine (D50 = 0,31 mm; U = 2,1; porosity of 45%) and coarse material (D50 = 1,4 mm; U = 1,72; porosity of 48%) and low dosages of oocysts (50–70per ml). The study of Logan was performed with unsaturated columns due to intermittent loading of the columns (simulating rainfall). In the filtrate of both columns oocysts were detected and removal varied between 2.7 and >4.5 log.

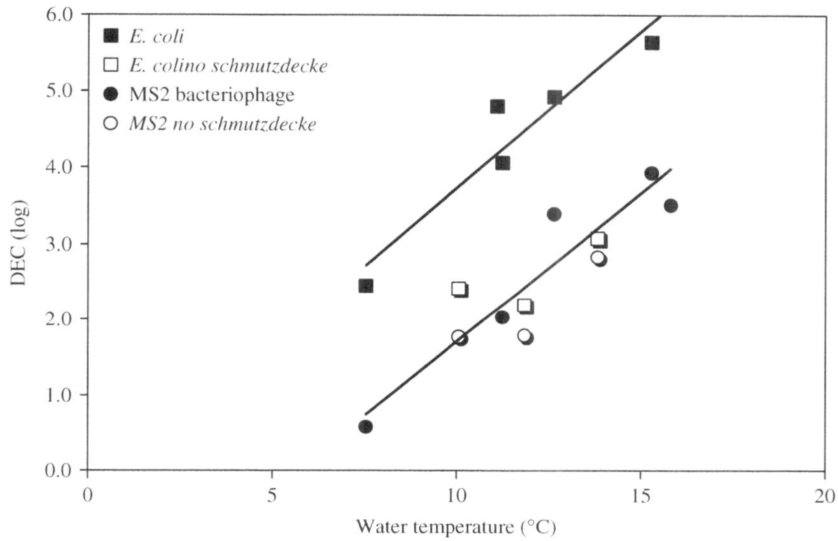

Figure 6.6 Influence of filter bed scraping and temperature on DEC of a slow sand filter

Although most of the column studies showed that peak concentrations of *Cryptosporidium* oocysts are removed very efficiently by sand filtration (Figure 6.4) other column studies with soil material revealed that oocysts can pass through relative fine material and demonstrated that besides straining, adsorption is also a mechanism responsible for removal:

- Nobel *et al.* (1999) for example showed a 3.6 log removal of oocysts in unsaturated natural soil columns while simultaneously spiked bacterial spores of *C. perfringens* were removed to a higher extent (5.2 log).
- Columns with fine sand (D50 = 0,18 mm; U = D60/D10 = 0,073; porosity of 36%) supplied with surface water enriched with oocysts of *C. parvum* removed these organisms with 3.6 and 3.2 log at an infiltration rate of 0.5 and 1.0 m/day, respectively (Hijnen *et al.*, 2005). In the same study more coarse but less uniform sand from the border of river Meuse (D50 = 0.5 mm; U = 5.3; porosity of 32%) supplied with river Meuse water enriched with oocysts removed these oocysts completely (>7 log). This indicates that beside straining, adsorption is also a removal mechanism for these pathogens.
- A delayed breakthrough of oocysts occurred after the dosage period of the experiment described by Hijnen *et al.* (2005); Figure 6.7). This peak coincided with a change in water quality. Especially a rise in pH (7.7 up to 8.3) may have increased electrostatic repulsion (increase of the negative charge density of the soil material and oocysts) and caused detachment. Detachment of oocysts was also observed in soil columns when the ion strength was reduced (Davies *et al.*, 2005).
- A breakthrough of *Cryptosporidium* oocysts in a full-scale and pilot plant SSF-filter was demonstrated by WRC (1995; 2.7 log) and Hijnen *et al.* (2007; 4.7 log).

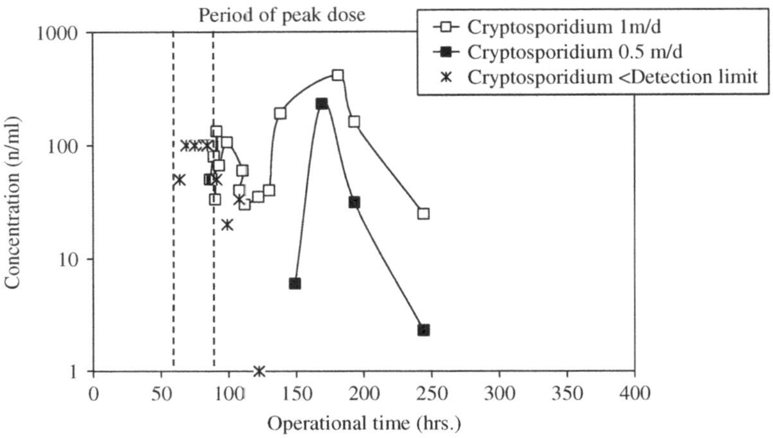

Figure 6.7 Concentration of oocysts observed in the filtrate of soil columns spiked during 24 hrs. with a concentration of 2000 oocysts per ml (from: Hijnen et al., 2005)

In the latter study, however, a significant decline of oocysts was observed and attributed to predation (micro faunal groups of organisms were detected in the filter bed described in literature as predators of oocysts). The large impact of temperature on the MS2 bacteriophage and *E. coli* is most likely also driven by biological activity in which predation may play a significant role.

6.4.2 Surrogates for pathogen removal in slow sand filtration

The MS2 phage used in most studies to determine DEC of SSF for viruses can be regarded as a solid surrogate for these pathogenic micro-organisms (Schijven, 2001). For pathogenic bacteria like *Campylobacter, E. coli* and Coli44 are safe surrogates to determine DEC of SSF.

A valid surrogate for *Cryptosporidium* and *Giardia* removal by SSF is still not available. Based on the results of the latest study (Hijnen *et al.*, 2007) it was concluded that environmental spores of sulphite-reducing clostridia (SSRC) and the small sized centric diatoms are too conservative parameters. This conclusion was based on the lower DEC-value of both surrogates compared to the DEC for *Cryptosporidium* and the observation that the numbers of oocysts retained in the filter bed declined as a result of predation. The latter observation minimises the risk of delayed breakthrough as observed for SSRC and the diatoms.

6.5 Conclusions

Slow sand filters are important barriers in water treatment for removal of pathogenic micro-organisms. The MEC of these filters for viruses and bacteria is 2–3 log, a value based on a reasonable amount of data. As expected *Cryptosporidium* and *Giardia* (oo)cysts are removed to a higher extent. The MEC-values of 4.8 and 4.9 log, respectively for both parasites, however, these values are based on a low number of data.

Process conditions like the presence of a schmutzdecke, hydraulic retention time, flow rate, grain size and temperature will affect elimination of micro-organisms. A stable operated filter bed covered by a roof, with a schmutzdecke and no channels (preferential flow) is a reliable and effective barrier in the treatment. Variation in the DEC will be minimised when filter bed cleaning (removal of schmutzdecke to restore production) is performed with hygienic precautions and these filters are put back to production after a filter-to-waste period. Finally, the data revealed that biological activity has a large impact on the DEC of slow sand filters. Predation by micro faunal groups most likely play a significant role in the removal process. This is one of the aspects which has to be elucidated in further research to increase the understanding about the removal of micro-organisms by slow sand filters.

7 Disinfection processes

As described in the methods, elimination of micro-organisms by disinfection processes cannot be evaluated in the same way as done for physicochemical removal processes, such as coagulation and slow sand filtration. The latter processes were evaluated as if all processes were operated in a general default mode. To differentiate between studies, differences in DEC were related to specific processes and/or study conditions. In most disinfection studies however, the objective is to determine a dose/inactivation relationship necessary to design and operate processes at a certain inactivation capacity. This "disinfection dose" is a combination of the concentration of the disinfectant C (or fluence rate in case of UV radiation) and the exposure time T. In this chapter the expected DEC for a certain a range of disinfection doses will be calculated from the evaluated studies.

7.1 UV-disinfection

The first application of UV irradiation in drinking water as disinfection process was in 1910 in Marseille (Henry *et al.*, 1910), after the development of the mercury vapour lamp and the quartz tube and establishing the germicidal effect of UV irradiation. According to Wolfe (1990) and Hoyer (2004) general application however was hampered because of high costs, poor equipment reliability, maintenance problems and the advent of chlorination (cheaper, more reliable and potential measurement of disinfectant residual). Due to the increased information on the production of hazardous oxidation by-products during chlorination and ozonation, UV irradiation gained more attention, due to the fact that UV produces almost no by-products. Also, unlike chemical disinfectants, the biological stability of the water is not affected by low pressure lamps. In Europe, UV has been widely applied for drinking water disinfection since the 1980's, for the control of incidental contamination of vulnerable groundwater and for reduction of Heterotrophic Plate Counts (Kruithof *et al.*, 1992). The breakthrough of UV applicability as a primary disinfection process in the US and Europe came after the discovery of the high efficacy of UV irradiation against *Cryptosporidium* (Clancy *et al.*, 1998) and *Giardia*. Chemical disinfection with chlorine is not effective against these pathogens and ozone (applied at low CT-values to limit formation of bromate) has relatively little effect on the infectivity of the protozoan oocysts. But infectivity of these pathogens is significantly reduced by UV fluences that can readily be applied in drinking water treatment. UV is now regarded as being broadly effective against all pathogens, including bacteria, protozoa and viruses, that can be transmitted through drinking water.

The introduction of the Quantitative Microbiological Risk Assessment (QMRA) to define the microbiological safety of drinking water (Haas, 1983; Regli *et al.*, 1991; Teunis *et al.*, 1997; Medema *et al.*, 2003) is a development of growing interest. Besides knowledge about the presence of pathogenic micro-organisms in the source water, QMRA requires quantitative knowledge about the capacity of water treatment processes, including UV disinfection, to eliminate (remove or inactivate) pathogenic micro-organisms.

The aim of this study was to evaluate available literature data and create a well-defined database which enables calculation of the Micro-organism Inactivation Credit (MIC) of UV disinfection for viruses, bacteria and protozoan (oo)cysts. Most studies are lab-scale based. For full-scale chemical disinfection

 Elimination of micro-organisms by Drinking Water Treatment Processes. By W.A.M. Hijnen and Gertjan J. Medema. ISBN: 9781843393733. Published by IWA Publishing, London, UK.

processes with chlorine and ozone, however, it was demonstrated previously that MIC for E. coli was lower than expected from the applied CT-values and the known dose-response curves determined under laboratory conditions (Hijnen *et al.*, 2000; 2004a).

Therefore literature was evaluated to verify the influence of process conditions on the MIC of full-scale UV-disinfection processes. The results of this review will also be used to identify gaps in current knowledge necessary to identify further research needs.

7.1.1 Viruses and bacteriophages

The number of studies where UV sensitivity of specific pathogenic viruses and bacteriophages is determined under well-defined laboratory conditions with collimated beam apparatus (CB-tests) ranged from 1 to 6. The total number of data per virus ranged from 3 up to 109 (Tables 7.1 and 7.2).

Table 7.1 Microbial Inactivation Credit (*MIC; log*) of monochromatic and polychromatic UV radiation for viruses determined with collimated beam tests

	Studies (n)	UV fluence (mJ/cm^2)	k[a] (±95%CI; r^2)	MIC_{max} (Log)
Poliovirus type 2	6 (61)[b,c,d,e,f,g,j]	5–50	0.135 (0.007; 0.79)	5.4
Adenovirus ST2,15,40,41	5 (98)[g,h,i,j,k]	8–306	0.024 (0.001; 0.87)	6.4
Adenovirus ST40	1 (29)[i]	8–184	0.018 (0.004; 0.88)	3.0
Adenovirus ST2,41 MP	1 (18)[k]	30–90	0.040 (0.003; 0.77)	4.3
Rotavirus SA-11	5 (55)[b,d,e,k,l]	5–50	0.102 (0.006; 0.78)	4.1
Rotavirus SA-11 MP	1 (11)[k]	5–30	0.154 (0.011; 0.92)	4.6
Calicivirus feline, canine	3 (29)[i,m,n]	4–49	0.106 (0.010; 0.67)	5.5
Calicivirus bovine	1 (20)[k]	4–33	0.190 (0.008; 0.96)	5.7
Calicivirus bovine MP	1 (20)[k]	2–15	0.293 (0.010; 0.97)	5.9
Hepatitis A	3 (13)[e,l,o]	5–28	0.181 (0.028; 0.70)	5.4
Coxsackie virus B5	2 (12)[h,l]	5–40	0.119 (0.006; 0.97)	4.8

[a]linear regression, intercept = 0; [b]Chang *et al.*, 1985; [c]Harris *et al.*, 1987; [d]Sommer *et al.*, 1989; [e]Wilson *et al.*, 1992; [f]Maier *et al.*, 1995; [g]Meng and Gerba, 1996; [h]Gerba *et al.*, 2002; [i]Thurston-Enriquez *et al.*, 2003; [j]Thompson *et al.*, 2003; [k]Malley *et al.*, 2004; [l]Battigelli *et al.*, 1993; [m]De Roda Husman *et al.*, 2003; [n]Duizer *et al.*, 2004; [o]Wiedenmann *et al.*, 1993

Table 7.2 Microbial Inactivation Credit (*MIC; log*) of monochromatic and polychromatic UV radiation for bacteriophages determined with collimated beam tests

	Studies (n)	UV fluence (mJ/cm^2)	k[a] (±95%CI; r^2)	MIC_{max} (Log)
MS2-phages	5 (109)[b,c,d,e,f]	5–139	0.055 (0.002; 0.93)	4.9
MS2-phages MP	1 (11)[f]	12–46	0.122 (0.009; 0.92)	5.3
φX174	**4 (30)[e,g,h,i]**	2–12	0.396 (0.025; 0.85)	4.0
PRD1	**1 (4)[d]**	9–35	0.128 (0.014; 0.98)	3.8
B40-8	**1 (14)[j]**	1–39	0.140 (0.010; 0.96)	5.6
T7	**1 (3)[j]**	5–20	0.232 (0.080; 0.90)	4.6
Qβ	**1 (5)[j]**	10–50	0.084 (0.003; 0.99)	4.2

[a] linear regression, intercept = 0; [b]Havelaar *et al.*, 1990; [c]Wilson *et al.*, 1992; [d]Meng and Gerba, 1996; [e]Sommer *et al.*, 1998; [f] Malley *et al.*, 2004; [g]Battigelli *et al.*, 1993; [h]Oppenheimer *et al.*, 1993; [i]Sommer *et al.*, 2001; [j]Clancy *et al.*, 2004

The calculated inactivation rate constant k (no shoulder; intercept = 0) showed a narrow 95% confidence-interval and a high goodness-of-fit (13 out of 17 organisms $r^2 > 0.85$). The six authors describing inactivation of seeded Poliovirus type 1 yielded a total of 61 data points presented in Figure 7.1. The inactivation rate constant k calculated for a UV fluence range of 5 to 50 mJ/cm^2 was 0.135 (95%-CI = 0.007; r^2 = 0.79). Due to the observed tailing by Sommer *et al.* (1989) and Maier *et al.* (1995), MIC_{max} is set at 5.4 log.

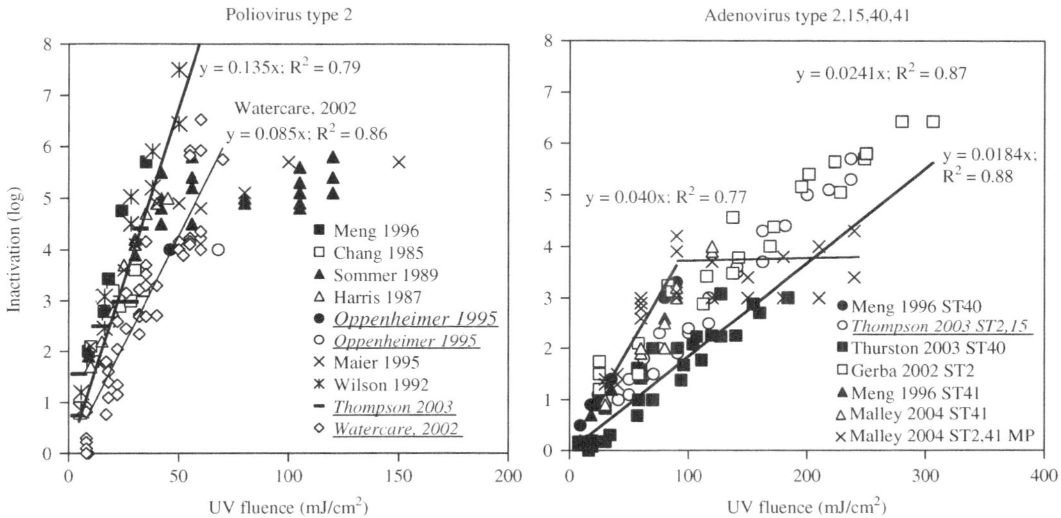

Figure 7.1 UV fluence-response curves for Poliovirus and Adenovirus (all DWS, but italic: WWS)

In Figure 7.1 the fluence-response curves for Adenoviruses serotypes (ST) 2, 15, 40 and 41 are also presented. These data demonstrate that the UV sensitivity of these serotypes for monochromatic UV radiation is almost similar. Adenovirus is the most persistent virus type presented in Table 7.1 and for Adenovirus ST40 tested by Thurston-Enriquez *et al.* (2003) the lowest k-value of 0.018 cm^2/mJ was determined. Malley *et al.* (2004) determined the UV sensitivity of Adenovirus ST2 and ST41 for polychromatic UV radiation (medium pressure lamps). Up to a UV fluence of 90 mJ/cm^2, the UV sensitivity was a factor of 1.7 higher than that observed for monochromatic UV radiation (Table 7.1), but above this fluence they observed tailing (Figure 7.1). By using bandpass filters they distinguished the germicidal effect of different wavelengths in the polychromatic UV light at fluence-ranges up to 90 mJ/cm^2 and showed that wavelengths of 220 and 228 nm UV were significantly more effective in inactivating Adenovirus ST2 than higher wavelengths.

The fluence-response data and lines for Rotavirus type SA11 and for three types of Caliciviruses are presented in Figure 7.2. Again Sommer *et al.* (1989) showed reduced inactivation at fluences above 50 mJ/cm^2 and Malley *et al.* (2004) showed that monochromatic UV radiation was less efficient than polychromatic UV radiation for inactivation of Rotavirus SA-11. For MP-lamps a 1.7 times higher sensitivity was calculated (Table 7.1). Caliciviruses from different non-human hosts (feline, canine and bovine) showed highest UV sensitivity for the bovine type (Malley *et al.*, 2004) and the same study showed a 1.5 times higher inactivation with polychromatic UV radiation (MP-lamps) compared to low-pressure lamps. The k-value for inactivation of Feline Calicivirus (0.106 cm^2/mJ) was in the same order of magnitude as that observed for Rotavirus, Poliovirus and Coxsackie virus B5 (Table 7.1). Hepatitis A virus was more sensitive to UV radiation.

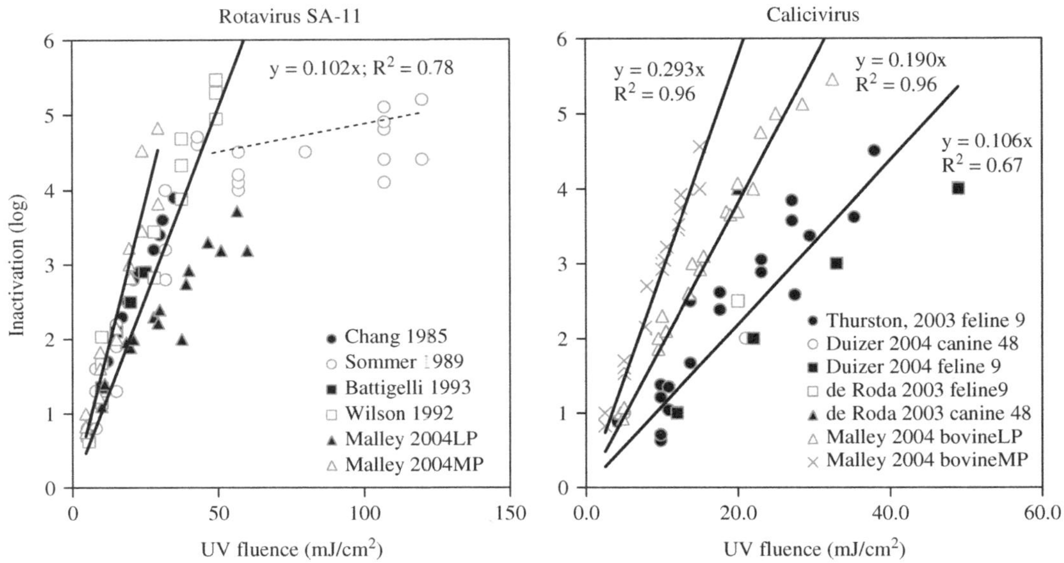

Figure 7.2 UV fluence-response curves for rotavirus and calicivirus

Noroviruses are part of the human caliciviruses and are not culturable. Using RT-PCR technique Watercare (2002) determined that Norovirus was relatively resistant to UV in comparison with other viruses. In a wastewater treatment plant, ten samples before and 10 samples after UV were analysed; seven samples before UV and one sample after UV were positive for Norovirus. The mean fluence was 23 mJ/cm^2. From these (presence/absence) data, a 0.8 log inactivation was estimated for a UV fluence of 20 mJ/cm^2. At higher fluences (40 and 70 mJ/cm^2) all samples were negative. However, it is uncertain to what degree inactivation Norovirus by UV assessed using RT-PCR is representative of reduction in infectivity of these viruses.

UV sensitivity of bacteriophages used or proposed as model organisms for the assessment of the REF of a UV system on full-scale are also presented in Table 7.2. MS2 phages are the most persistent of the tested phages with a k-value of approximately two times the k-value of the most persistent Adenovirus (Table 7.1). k-Values of the other bacteriophages ranged from 0.128 cm^2/mJ for PRD1 up to 0.396 cm^2/mJ for PhiX174.

As shown for Rotavirus and Calicivirus, polychromatic radiation (MP-lamps) was more effective, by a factor of 2.2, than monochromatic UV radiation according to the study of Malley *et al.* (2004). UV fluence of the MP-lamp was calculated based on the average irradiance measured by a UV-sensor and weighted by a germicidal factor at each wavelength (based on the DNA absorbance, relative to 254 nm). Thus, the fluence of LP- and MP-lamps were compensated for the wavelengths emitted by these lamp. Malley *et al.* (2004) argued that this weighting may have been biased for MP-lamps. On the other hand their results may indicate a higher inactivation efficiency of MP lamps compared to LP lamps. Remarkable though, was the observation that for all tested viruses and phages (adeno-, rota-, caliciviruses and MS2 phages) MP lamps were approximately 2 times more effective than LP lamps. Assuming a more variable sensitivity of these different viruses to polychromatic UV radiation, incorrect or biased weighting of the fluence of the MP lamps seems the most plausible cause for the observed higher efficiency of these lamps compared to LP lamps.

7.1.2 Bacteria and bacterial spores

Bacteria (vegetative cells) are significantly more susceptible to UV radiation than viruses and therefore less extensively studied. In Figure 7.3, fluence-response curves for some selected pathogenic bacteria are presented. With the exception of *E. coli* in five studies, only one or two studies were found for specific pathogenic bacteria.

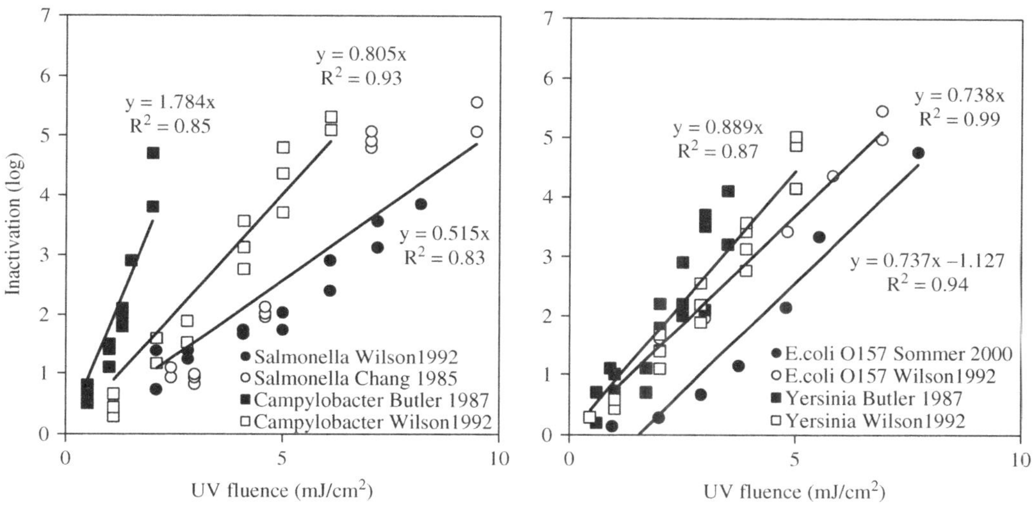

Figure 7.3 UV fluence-response curves for pathogenic bacteria

Wilson *et al.* (1992) tested the UV sensitivity of seven of the ten bacterial species presented in Table 7.3. The number of data points ranged from 4 up to 37 for *E. coli*. The k-values varied from 0.312 cm^2/mJ for *Streptococcus faecalis* to 1.341 cm^2/mJ for *Vibrio cholerae* (both Wilson *et al.*, 1992). Linear regression analysis showed low variation (95% confidence interval) and high goodness-of-fit (r^2).

Aerobic spores of *Bacillus subtilis* and anaerobic spores of *Clostridium perfringens* are more resistant against UV. The UV sensitivity of the latter spore-forming bacterium was determined in a continuous flow-system with medium pressure lamps with UV fluences assessed with biodosimetry (REF).

7.1.3 Pathogenic protozoa

Interest in UV as a disinfection process for water has increased after Clancy *et al.* (1998) showed that *Cryptosporidium parvum* oocysts were highly susceptible to UV when the effect on the infectivity was assessed with the neonatal mouse model. Since then, Clancy and several other authors have studied inactivation of *Cryptosporidium parvum* and *Giardia muris* by UV radiation (Table 7.4).

Figures 7.4 and 7.5 show substantial inactivation of (oo)cysts of both protozoa at low UV fluences (<20 mJ/cm^2) by low- and medium-pressure lamps. Recently, Johnson *et al.* (2005) demonstrated a similar UV sensitivity for *C. hominis* oocysts which predominates in human cryptosporidiosis infections. Based on the regression analysis of these fluence-response data, efficacy of low- and medium-pressure lamps for oocyst inactivation is in the same order of magnitude (Table 7.3). Comparison of these k-values with the k-values from Tables 7.1 and 7.2 show that these protozoa are more sensitive to UV than viruses, but less sensitive compared to most bacteria. The regression analysis of the accumulated data shows a low goodness-of-fit ($r^2 = -0.99$–0.49) and a positive intercept value. Furthermore, Craik *et al.* (2000, 2001) observed considerable tailing for a number of inactivation data at high UV fluences (Figure 7.4). Qian *et al.* (2004) described these data with a statistical method (Bayesian meta-analysis) which resulted in the UV fluence requirement curves presented in Figures 7.4 and 7.5 (Pirnie *et al.*, 2003). These curves were calculated for an inactivation requirement of up to 3 log and could be described by a log-log relationship (Log inactivation of *Giardia* = 1.2085 LN UV fluence + 0.0715; $r^2 = 0.99$; Log inactivation of *Cryptosporidium* = 1.2344 LN UV fluence–0.1283; $r^2 = 0.99$).

Table 7.3 Microbial Inactivation Credit (MIC; log) of monochromatic and polychromatic UV radiation for bacteria and bacterial spores determined with collimated beam tests

	Studies (data)	Fluence (mJ/cm^2)	k (±95%CI; r^2)	Offset[a] (mJ/cm^2)	MIC_{max} (Log)
Salmonella typhi	2 (26)[b,c]	2–10	0.515 (0.047; 0.83)	0	5.6
Campylobacter jejuni[a]	2 (27)[c,d]	0.5–6	0.880 (0.124; 0.65)	0	5.3
Yersinia enterocolitica	2 (34)[c,d]	0.6–5	0.889 (0.060; 0.87)	0	5.0
Shigella dysenteriae	1 (9)[c]	1–5	1.100 (0.096; 0.99)	0.69	5.9
Shigella sonnei	1 (9)[b]	3–8	0.468 (0.053; 0.89)	0	4.7
Vibrio cholerae	1 (10)[c]	0.6–4	1.341 (0.113; 0.94)	0	5.8
Legionella pneumophila	1(15)[c]	1–12	0.400 (0.040; 0.92)	0	4.4
Legionella pneumophila	1 (4)[e]	0.5–3	1.079 (0.077; 0.99)	0	3.0
E. coli O157[a]	2 (16)[c,f]	1–7	0.642 (0.082; 0.85)	0	5.5
E. coli	6 (41)[b,d,g,h,i,j]	1–15	0.506 (0.049; 0.71)	0	6.0
Streptococcus faecalis	2 (19)[b,g]	2.5–16	0.312 (0.032; 0.85)	0	4.6
Bacillus subtilis	4 (30)[b,h,l,k]	5–78	0.059 (0.007 0.91)	−0.738	4.0
C. perfringens MP	1 (9)[l]	48–64	0.060 (0.027; 0.81)	−1.077	3.0

[a]Average of two spp.; Offset is threshold-value > 0: linear regression with intercept ≠ 0; [b]Chang *et al.*, 1985; [c]Wilson *et al.*, 1992; [d]Butler *et al.*, 1987; [e]Antopol *et al.*, 1979; [f]Sommer *et al.*, 2000a; [g]Harris *et al.*, 1987; [h,i] Sommer *et al.*, 1989, 1996a; [j]Zimmer *et al.*, 2002; [k]Sommer *et al.*, 1998; [l]Hijnen *et al.*, 2004b (continuous flow system)

Table 7.4 Microbial Inactivation Credit (MIC; log) of monochromatic and polychromatic UV radiation for protozoa and Acanthamoeba spp. determined with collimated beam tests

	Studies (data)	k (±95% CI; r^2)	UV fluence (mJ/cm^2)		MIC_{max}
			Range	intercept (95%)	
C. parvum LP	6 (38)[a,b,c,d,e,f]	0.243 (0.08; 0.49)	0.5–6.1	1.502	3.0
C. parvum MP	4 (65)[a,c,f,g]	0.225 (0.07; 0.37)	0.9–13.1	1.087	
Giardia muris	1 (4)[h]	0.282 (0.164; −0.99)	1.5–11	0	2.4
Giardia lamblia	1 (2)[i]	nd	0.05–1.5	nd	2.5
Acanthamoeba spp.	1 (16)[j]	0.021 (0.004; 0.94)	43–172	0.499 (0.449)	4.5

[a,b]Clancy *et al.*, 2000, 2002; [c]Craik *et al.*, 2001; [d]Shin *et al.*, 2001; [e]Morita *et al.*, 2002; [f]Rochelle *et al.*, 2004; [g]Bolton *et al.*, 1998; [h]Craik *et al.*, 2000; [i]Linden *et al.*, 2002; [j]Maya *et al.*, 2003

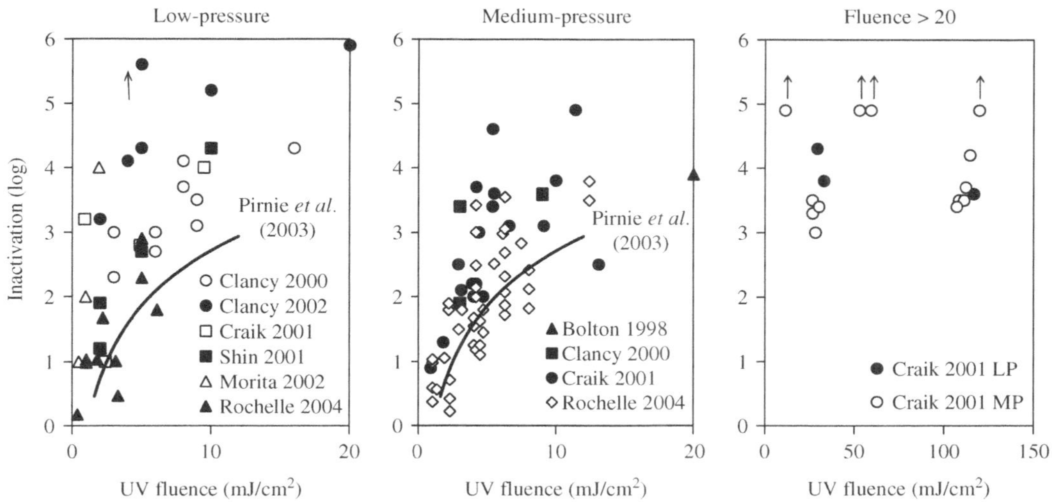

Figure 7.4 UV fluence-response curves for Cryptosporidium parvum (multiple strains) by medium-pressure (a) and low-pressure (b) UV lamps

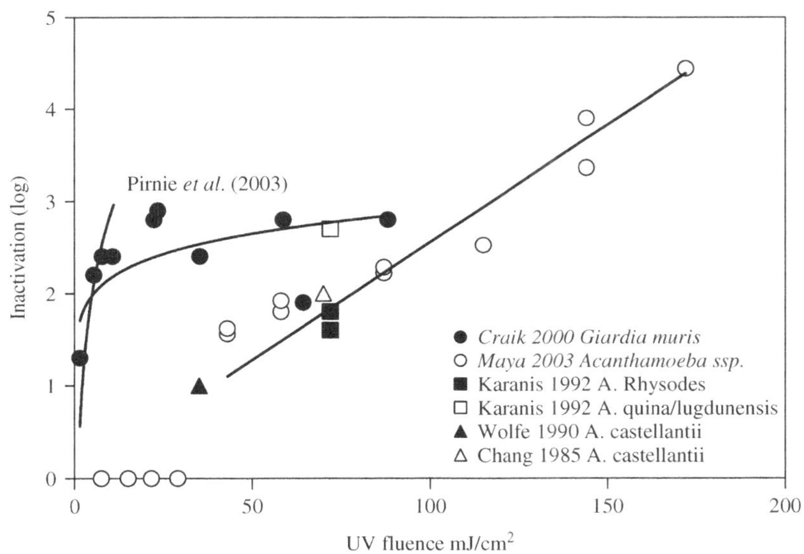

Figure 7.5 UV fluence-response curves for Giardia muris and Acanthamoeba spp.

One study described the inactivation of *Acanthamoeba spp.* by UV in CB-tests using CD1 neonatal mouse model test to measure infectivity (Maya *et al.*, 2003). Just as observed for *B. subtilis* a threshold UV fluence is required for this organism to see an effect on infectivity of these pathogens. This threshold value of 30 mJ/cm^2 as well as the k-value of 0.021 calculated from the successive log-linear relationship, show that the sensitivity of this micro-organism and of the most resistant virus type Adenovirus to UV is in the same order of magnitude.

7.1.4 Discussion

The k-values summarised in Tables 7.1, 7.2 and 7.3 can be used to design new processes or to calculate the inactivation efficiency of operational UV systems. Translation of UV sensitivity assessed with CB-tests and seeded micro-organisms to the efficiency of UV disinfection under full-scale conditions, however, is influenced by factors related to the micro-organisms and by factors related to the fluence assessment. For chemical disinfection with ozone this was demonstrated recently by Smeets *et al.* (2004; 2005). The literature on the influence of these factors has been reviewed. For pathogenic micro-organisms two studies were found (Figure 7.6) that investigated the inactivation of environmental pathogenic micro-organisms under different conditions. For indicators, more studies were available. Most of the evaluated data came from wastewater studies and to a lesser extent from drinking water studies. Inactivation of seeded or environmental indicator micro-organisms (coliforms, enterococci, clostridia spores, FRNA-phages, *Bacillus* spores) has been determined in either CB-apparatus or CF-systems. Inactivation data are presented in Figures 7.7–7.11, where CB-test results are separated from results from CF-systems. The findings are reviewed in the two following paragraphs.

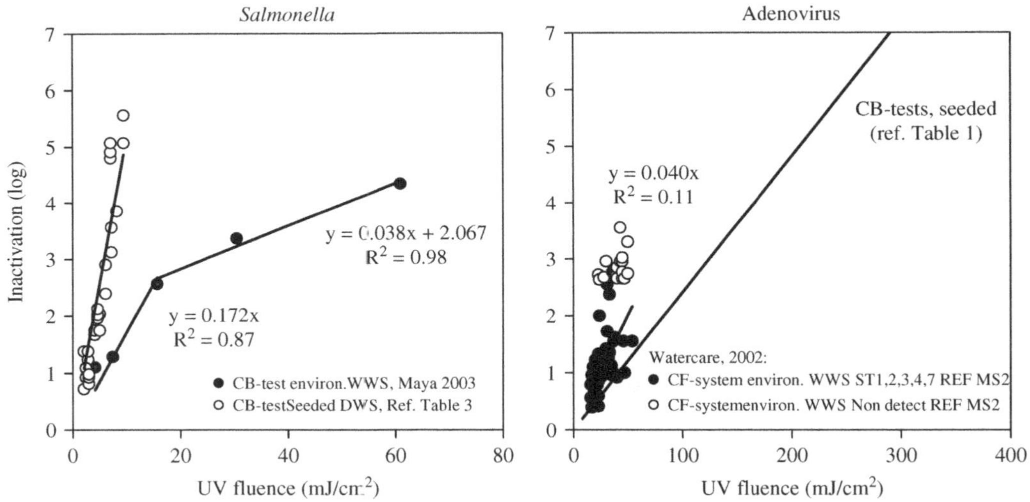

Figure 7.6 Comparison of UV fluence-response curves for seeded and environmental Salmonella bacteria and Adenoviruses; separate studies under different conditions

***Micro-organism related factors*.** The UV sensitivity of seeded and environmental micro-organisms is compared in some studies under identical conditions or by comparing results from a study with environmental organisms with the overall data for seeded organisms tested in CB-apparatus or CF-systems (DWS, Figures 7.6–7.9). Environmental *Salmonella*, faecal coliforms and enterococci in CB-tests in wastewater (Maya *et al.*, 2003) were more resistant to UV light than the seeded micro-organisms of the same species (DWS, Figures 7.6, 7.8 and 7.9). A higher UV resistance of environmental spores compared to seeded spores (which were surviving isolates from the environmental spores) was also observed for *Bacillus* spp. (Mamane-Gravetz *et al.*, 2005; CB-tests) and sulphite-reducing clostridia (Hijnen *et al.*, 2004b; CF-system), both DWS (Figure 7.10).

In the WWS-study of Watercare (Watercare, 2002; Simpson *et al.*, 2003; Jacangelo *et al.*, 2004), however, the UV sensitivity of environmental F-specific RNA phages (FRNA) was comparable to the UV sensitivity of seeded MS2 phages tested under similar conditions (CB-tests; Figure 7.7). A k-value of 0.049 cm^2/mJ was calculated for the seeded MS2-phages while for environmental FRNA tested under similar conditions in CB-tests resulted in a k-value of 0.047 cm^2/mJ (Figure 7.7).

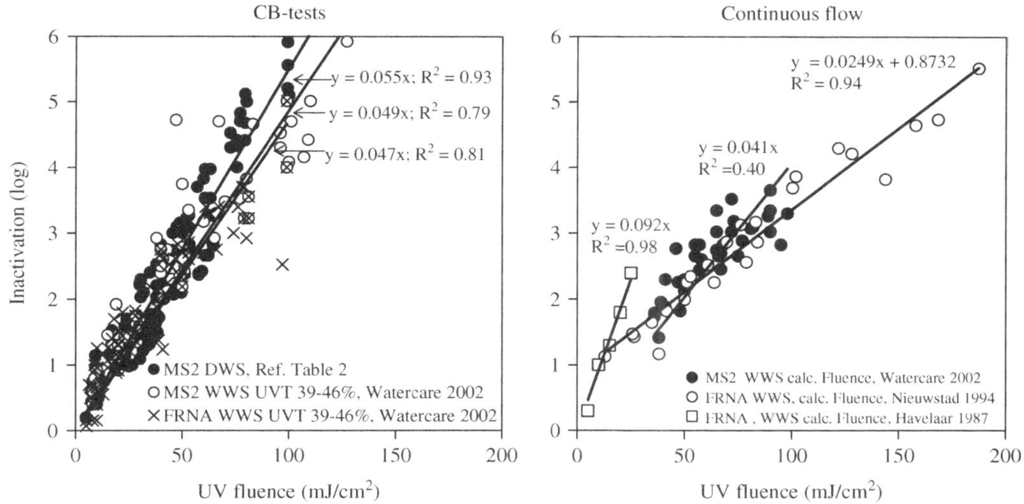

Figure 7.7 UV fluence-response curves for seeded MS2 FRNA phages and environmental FRNA phages determined under different conditions

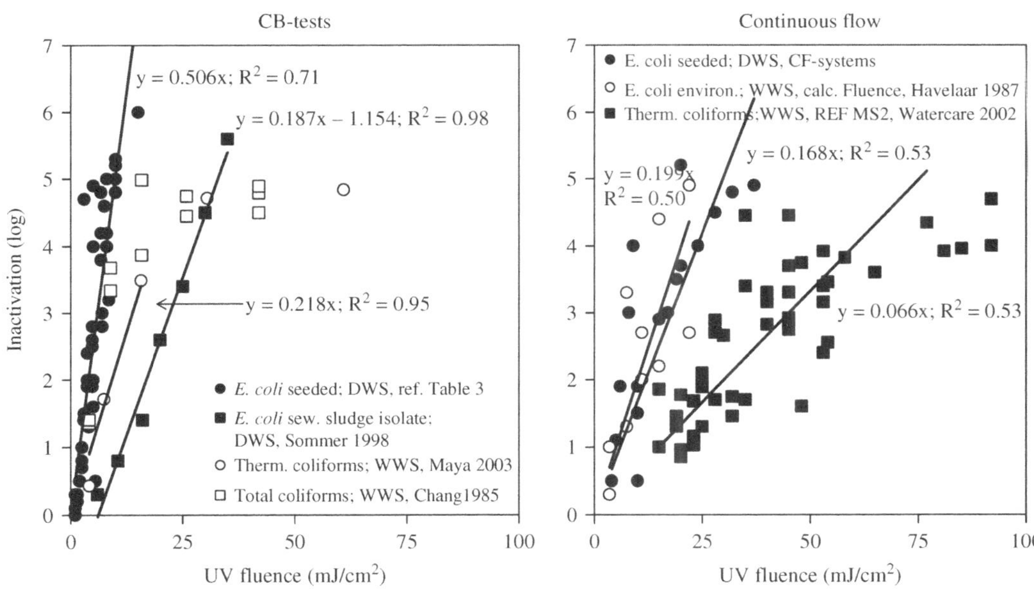

Figure 7.8 UV fluence-response curves for seeded and environmental coliforms (E. coli, thermotolerant coliforms and total coliforms) determined under different conditions

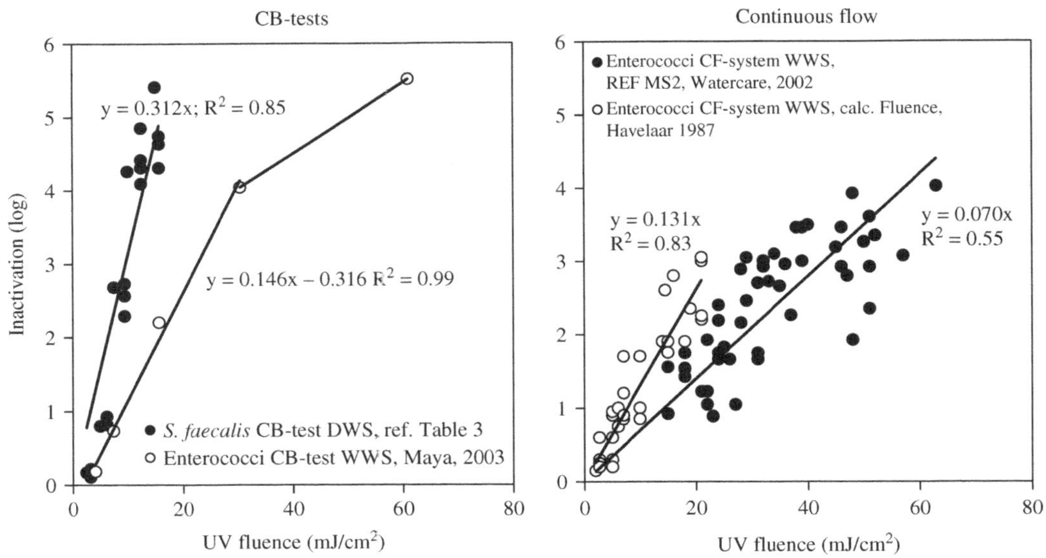

Figure 7.9 UV fluence-response curves for seeded and environmental enterococci determined under different conditions

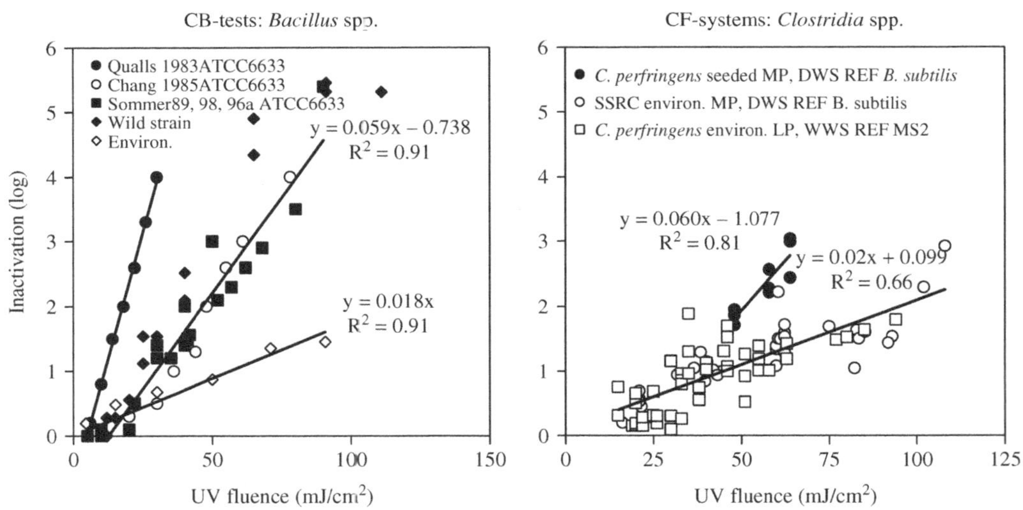

Figure 7.10 UV fluence-response curves for seeded and environmental bacterial spores determined under different conditions

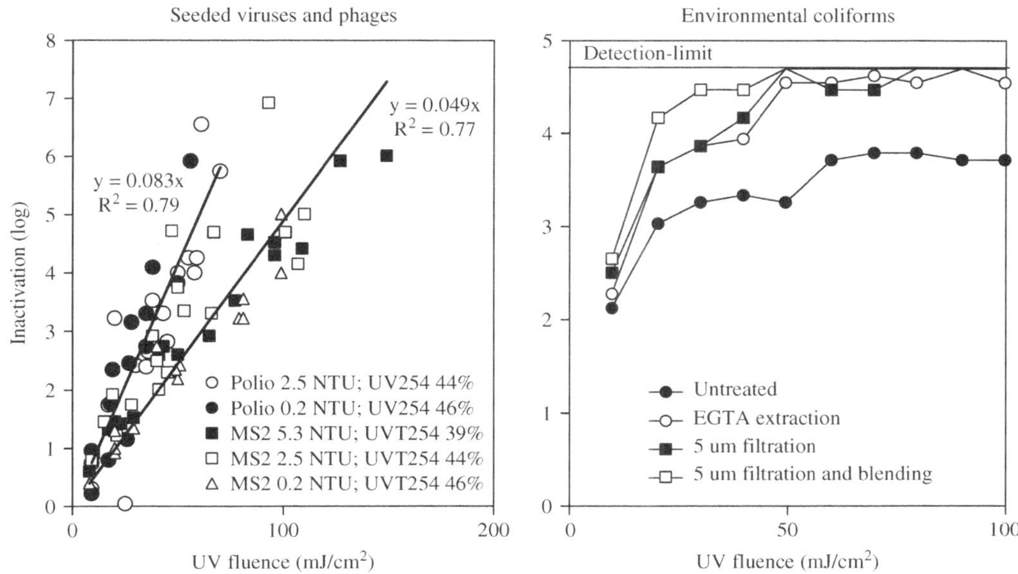

Figure 7.11 Effect of water quality and particle association on UV fluence-response curves for viruses and phages and for environmental coliforms, respectively

In contrast, a lower UV resistance of environmental Adenoviruses was observed in a CF-system (Watercare study) compared to seeded Adenoviruses tested in DWS-studies (CB-tests, Figure 7.6). Because 35% of the observations in this study yielded a minimal inactivation of >2.7 up to >3.3 log at a fluence range of 23 up to 50 mJ/cm^2 (Figure 7.6), the difference is even larger; these data have not been used in the k-value calculation presented in Figure 7.6. Predominant Adenovirus types in the Watercare study were serotypes 1, 2, 3, 4 and 7, with less commonly serotypes 5, 8, 11, 13, 15, 19, 25 and 29. The higher susceptibility of the environmental Adenovirus in this study could be the result of the absence of Adenovirus, type 40, the most persistent serotype. For (oo)cysts of *Cryptosporidium* and *Giardia*, no information is available on the UV sensitivity of environmental isolates.

Overall, increased UV resistance of environmental micro-organisms was more explicit for the bacterial spores than for vegetative bacteria and was of less significance for FRNA-phages and viruses.

A. Physiological state. This observed difference in UV sensitivity may be attributable to the physiological state of the micro-organisms, strain diversity, DNA-repair mechanisms and particle association. The physiological state of micro-organisms affects the sensitivity to environmental stress factors such as UV radiation. Martiny *et al.* (1990), Mofidi *et al.* (2002) and Malley *et al.* (2004) showed that UV sensitivity was related to the growth-phase of the bacteria; with the highest sensitivity in the active growth phase and lower sensitivity in the stationary phase.

B. Strain variation. Different strains of one species may have different UV sensitivity, as demonstrated for *E. coli* by Sommer *et al.*, 2000a, Sommer *et al.*, 1998 (Figure 7.8) and Malley *et al.* (2004). UV sensitivity of different *E. coli* strains in these studies varied by a factor of 5.8 and 3.7, respectively. The latter study demonstrated a higher sensitivity of *E. coli* O157:H7 compared to non-pathogenic/toxic strains. In contrast, Clancy *et al.* (2002) and Rochelle *et al.* (2004) showed that the high inactivation efficiency of UV radiation for *Cryptosporidium* was observed in multiple strains of *C. parvum*. The similar UV sensitivity observed for *C. hominis* (Johnson *et al.*, 2005) suggests that this high sensitivity of *Cryptosporidium* oocysts is common for all sub-species.

C. Repair. Because micro-organisms that have been exposed to UV light still retain metabolic functions, and in organisms that have been exposed to UV from the sun, DNA repair mechanisms have evolved over time, organisms are able to repair damage and regain infectivity. Photo-reactivation or dark-repair may occur. Photo-reactivation is restricted to conditions of prolonged exposure to light. The mechanisms of repair are described in von Sonntag *et al.* (2004). Dark-repair has been demonstrated in many bacteria. Spores have no active metabolism, but repair starts upon germination. Several viruses have been shown to use the repair enzymes of the host cell. This is suggested as the cause of the high resistance of Adenovirus, a double-stranded DNA virus, which can use the host cell's repair mechanism, while RNA-viruses may not. Some viruses even carry the genes for repair-enzymes (Lytle, 1971; refs. von Sonntag). This has not been demonstrated for the viruses that are transmitted via water, however.

The repair mechanism employed depends on the type of micro-organism, degree of inactivation (level of UV fluence), time between exposure and reactivation and the nutrient state. For bacteria and DNA viruses some studies demonstrated repair after exposure. Quantitative data showed a 6 times higher UV fluence requirement for 3 log inactivation of *Legionella* species (Knudson, 1985; Oguma *et al.*, 2002a). Bernhard (1994) described increased fluence requirement after photo-reactivation for *E. coli* (1.8–4 times), *Yersinia enterocolitica* (3.3 times), *Salmonella typhi* (1.4 times) and *Vibrio cholera* (4.2 times). For *E. coli* similar enhancement of fluence requirement was observed by Hoyer (1998; 3.5 times for 4 log inactivation) and Sommer *et al.* (2000a; 2.5 times) after photo-reactivation. The latter author showed hardly any dark-repair for *E. coli*. This observation was confirmed by Zimmer *et al.* (2002), who demonstrated that photo-reactivation did not occur after MP-UV, an observation also supported by Oguma *et al.* (2002b). Morita *et al.* (2002) demonstrated photo-reactivation and dark-repair of DNA in *Cryptosporidium parvum* with the endonuclease-sensitivity site assay. The animal infectivity, however, was not restored. Furthermore, they concluded that UV radiated oocysts are able to excyst but have lost their ability to infect host cells. Similar observations were reported by Shin *et al.* (2001) and Zimmer *et al.* (2003). Craik *et al.* (2000) and Linden *et al.* (2002) came to the same conclusion for *Giardia muris* and *Giardia lamblia* cysts, respectively. Belosevic *et al.* (2001), however, showed the ability of DNA-repair by some *Giardia* spp. after UV radiation with medium pressure lamps. This was supported by the results presented by Kruithof *et al.* (2005) who showed invivo reactivation (dark repair) of *G. muris* cysts at fluence values as low as 25 mJ/cm^2 MP UV, but not at 60 mJ/cm^2, after prolonged time of incubation (3, 14 and 20% reactivation after 10, 20 and 30 days, respectively). Also DNA-repair of *G. lamblia* cysts after exposure to monochromatic UV irradiation is recently reported (Shin *et al.*, 2005). An extensive study was presented on repair in *C. parvum* oocysts by Rochelle *et al.*, in 2004. Identification of possible DNA repair genes in *C. parvum* showed that the oocysts contain all of the major genetic components of the nucleotide excision repair complex. Nevertheless, inactivation displayed by oocysts immediately after UV exposure or displayed by oocysts after UV exposure followed by various repair conditions were generally in the same order of magnitude. This demonstrates that repair of UV induced damage in *C. parvum* after UV exposure in drinking water is not likely to occur.

D. Particle association. Higher resistance of particle associated faecal bacteria has been observed in secondary effluents. This phenomenon was demonstrated by Qualls *et al.* (1983b) and Havelaar *et al.* (1987) for thermotolerant coliforms and the enterococci, respectively. A lower inactivation rate was observed in non-filtered effluent of sewage water plants compared to filtered samples (pore size 8 μm). Recently, Örmeci and Linden (2002) applied different techniques (extraction with EGTA, filtration of 5 μm filters with or without homogenization by blending) to separate particle- and non-particle associated coliforms and showed an increased resistance of environmental coliforms associated with particles to UV (Figure 7.11). Aggregates of *B. subtilis* spores were artificially made with clay in a Jar Test apparatus by Mamane-Gravetz and Linden (2004) and caused a reduction in inactivation efficiency. The k-value decreased from 0.0617 for the suspended spore-clay solution to 0.0579 cm^2/mJ for the aggregated spore-clay suspension. The same authors published a new study in 2005, in which they found evidence for a correlation of hydrophobicity of spores with aggregation. Aggregation may be a cause for tailing (lower inactivation at higher fluence) observed in the kinetics. The k-values of isolated environmental *Bacillus* strains in the tailing phase of the kinetics were similar to the k-values of the original and natural *Bacillus* spore population. This indicates that a shielding effect of aggregation or particle association is a significant factor in determining the low susceptibility of environmental spores also observed for the environmental clostridial spores as presented in Figure 7.10 (Hijnen *et al.*, 2004b).

Fluence related factors. Variability in fluence may be caused by water quality (adsorption, reflection and refraction) and the distribution of the hydraulic retention time in continuous flow systems.

A. Water quality. By placing 0.22 μm membrane filtered secondary effluent (UV transmission of 40–60%) in the UV pathway of the collimated beam apparatus, Havelaar *et al.* (1990) showed no decrease of the inactivation rate constant for MS2 phages. This illustrates that fluence values can easily be corrected for the presence of UV absorbing organic matter in the water. In studies with CF-systems both Schoenen *et al.* (1995) and Sommer *et al.* (1997) showed with biodosimetry that the influence of the lamp intensity upon the inactivation efficiency is higher than reduction of the water transmittance. In addition, the latter author also demonstrated that this difference was less defined in reactors with multiple lamps.

The influence of water quality on the efficiency of UV disinfection can be demonstrated by comparing results of drinking water studies (DWS) with wastewater studies (WWS) conducted under similar conditions. CB-tests with seeded Polioviruses in secondary effluent with high turbidity and low UV transmission were published by Oppenheimer *et al.* (1995), Watercare (2002) and Thompson *et al.* (2003) (Figure 7.1). In the first two studies a lower inactivation was observed compared to the DWS results. In the Watercare study inactivation rate constant k of seeded Poliovirus in filtered secondary effluent was a factor 1.6 lower than k-value assessed in drinking water studies (Figure 7.1). In the same study (Watercare, 2002), a decreased inactivation of MS2 compared to the accumulated drinking water data was also observed, but this decrease was minimal (a factor of 1.1; Figure 7.7). CB-tests were conducted with secondary effluent with high and low turbidity attained by further filtration (0.2–2.5 NTU and UVT of 40–68%). They clearly demonstrated that there was no impact of turbidity on the inactivation of seeded Polioviruses and MS2 phages (Figure 7.11). In addition data from the WWS-study of Thompson *et al.* (2003) showed no decreased inactivation of seeded Poliovirus and Adenovirus compared to inactivation of these organisms tested in DWS-studies (Figure 7.1).

B. Fluence determination in CF-systems. In the literature, only few drinking water studies have been published where fluence-response curves were determined with continuous flow systems (CF-systems). Results from the studies of Martini *et al.* (1990), Schoenen *et al.* (1991), Bernhardt *et al.* (1992, 1994) where fluence data were based on supplier of the equipment or on actinometry, showed lower inactivation rate constants for *E. coli* (Figure 7.8) when compared to the k-value determined from CB-tests in DWS.

Spores of *Bacillus subtilis* and MS2 phages are used as model organisms in biodosimetry assays for the assessment of the Reduction Equivalent Fluence (REF) of CF-systems. Sommer *et al.* (2000b) determined the REF of more than 30 commercial available CF-systems with UV_{254} calibrated spores of *B. subtilis* and presented the results of six systems. In one system, the REF was equal to the UV fluence calculated according to the supplier, four systems showed that that REF was 19–38% lower than the calculated fluence and in one system REF was 33% higher than the calculated fluence (Table 7.5).

The overestimation of fluence by calculation was supported by data presented by Hijnen *et al.* (2004b) and Watercare (2002). In the latter study REF of the CF-systems was determined with MS2 phages (Figure 7.7). Inactivation results showed high variability and the average ratio of the REF with the calculated fluence was 0.83 with a relatively high standard deviation of 0.25 (Table 7.5). The higher variability might be due to the higher turbidity and/or lower UV transmission of the water compared to the variability observed in the drinking water studies (Table 7.5). These observations show that assessing efficiency of UV disinfection based on calculated fluence data may overestimate and sometimes underestimate the MIC of the CF-system. It is the authors experience that the improvement in the fluence calculation by the introduction of Computational Fluid Dynamics and the recognition of the discrepancy between the REF and the calculated fluence has led to an improvement in the correspondence between calculated fluence and REF in recent years.

Table 7.5 Calculated UV fluence versus fluence assessed with biodosimetry (*REF*)

System	Study	Model organism	n	Ratio $UV_{calculated}$/REF (±SD)
A[a]	DWS	*B. subtilis*	3	1.33 (0.07)
B[a]	DWS	*B. subtilis*	3	0.81 (0.02)
C[a]	DWS	*B. subtilis*	3	0.79 (0.10)
D[a]	DWS	*B. subtilis*	3	1.00 (0.28)
E[a]	DWS	*B. subtilis*	3	0.73 (0.19)
F[a]	DWS	*B. subtilis*	3	0.62 (0.14)
G[b]	DWS	*B. subtilis*	9	0.59 (0.03)
H[c]	WWS	*MS2 phage*	37	0.83 (0.25)

[a]Sommer *et al.*, 2000b; [b]Hijnen *et al.*, 2004b; [c]Watercare, 2002

C. Reflection. Reflection caused by the construction materials of the UV reactor will have an influence on the inactivation efficiency determined by biodosimetry (Sommer *et al.*, 1996b). This factor is of greater influence in one lamp systems than in multiple lamp systems because of the higher surface-volume ratio.

7.1.5 General discussion

Kinetics of UV inactivation. Most of the inactivation data can be adequately described with the first-order disinfection model, at least for a certain fluence range. A threshold UV fluence before inactivation occurs, i.e., a shoulder model, is observed to some extent for bacteria but is more pronounced for *Bacillus* spores and *Acanthamoeba* spp. We have used the simple inactivation model where the shoulder is given as an offset of the first-order model.

Reduced inactivation at higher UV fluences, i.e., tailing, is observed in several drinking water studies with CB-tests (Polioviruses, rotaviruses, MS2 phages, *E. coli, C. parvum* and *G. muris*) and also for environmental bacteriophages and bacteria in wastewater studies in CF-systems. Tailing normally starts after at least 99% of the initial available micro-organisms are inactivated and is observed to a higher extent in the more UV susceptible micro-organisms. For the most resistant organisms (Adenoviruses, bacterial spores and *Acanthamoeba* spp.) tailing was not observed. The cause of tailing is still under debate. Several causes have been hypothesised, such as experimental bias, aggregation of micro-organisms, a resistant subpopulation or hydraulics, but no conclusive evidence is available for any of these. For micro-organisms where tailing is observed, we have used the first-order model only for the fluence range that yielded a linear relation with the inactivation. Because of the tailing, these kinetics cannot be extrapolated to higher fluences. For use in QMRA, the higher fluences can be assumed to yield (at least) the same inactivation credits as the highest fluence in the linear relation.

Significance for water disinfection. An extensive overview of the efficacy of UV disinfection for viruses, bacteria and bacterial spores and protozoan (oo)cysts was obtained from the reviewed literature. The k-values that were calculated from the reviewed studies can be used in QMRA and treatment design to determine the efficacy of a UV fluence in the inactivation of the range of reported bacterial and viral pathogens and indicator organisms. For *Cryptosporidium* and *Giardia*, the logarithmic functions given in Pirnie *et al.* (2003) were used for calculating the inactivation efficacy.

Of the pathogens, viruses are generally more resistant than *Cryptosporidium, Giardia* and the bacterial pathogens. Adenovirus 40 is the most UV resistant waterborne pathogen known. *Acanthamoeba* is also very resistant. Bacterial spores, esp. environmental spores of *Clostridium* are also resistant to UV, with k-values that are comparable to Adenovirus 40.

Correction: micro-organism related factors. Data from the evaluated studies indicated that photo-reactivation is more prominent and effective in repair of UV damage to (bacterial) DNA than dark-repair. Consequently, correction of the required fluence for DNA-repair is of significance for UV disinfection of wastewater, but not for UV disinfection of drinking water, where photo-reactivation under influence of light is not likely to occur. Hence for application in drinking water, no correction of the MIC for repair is needed. For application in wastewater, a correction factor of 4 is suggested to encompass the possibility of photo-reactivation of enteric bacteria, based on the available literature. For viruses, it is assumed that repair is included in the available fluence-response curves, as suggested for the double-stranded DNA Adenoviruses. Most studies on DNA repair of protozoan (oo)cysts indicate that correction for repair is not needed. However, dark-repair was observed in a recent study for *Giardia* cysts at UV fluences of 20 mJ/cm^2.

Based on the increased UV resistance observed for environmental *Salmonella*, enterococci, FRNA phages and spores of sulphite-reducing clostridia, correction of the fluence requirement for inactivation of bacteria and bacterial spores from the environment seems appropriate. The evaluated studies suggests a two times increased fluence requirement for bacteria and four times for bacterial spores. The results of environmental FRNA phages and Adenoviruses of the Watercare study indicate that such a correction is not needed for phages and viruses. However, further research is needed to support these findings. Similarly, studies on the increased resistance of environmental protozoan (oo)cysts to UV are appropriate.

Correction: fluence related factors. Most of the studies that have been reviewed have been executed under well-controlled laboratory conditions in which UV fluence was assessed with sensors and seeded micro-organisms. Information about the efficacy of UV systems under full-scale conditions was limited and those which have been evaluated, generally showed lower inactivation efficiency than in the laboratory. This reduced efficiency may be caused by factors related to the micro-organisms as described previously, but also by imperfections in the calculation of the fluence to which the micro-organisms are exposed in full-scale UV systems. The latter can be largely overcome by applying biodosimetry to full-scale UV systems to determine the Reduction Equivalent Fluence (REF). This is already enforced for the application of UV systems in drinking water practice in Austria (Österreichisches normungsinstitut, 1999). In Germany, a similar protocol is used as guideline (DVGW, 1997), and in the USA, the draft EPA Ultraviolet Disinfection Guidance Manual also appoints credits for inactivation of *Cryptosporidium* on the basis of biodosimetry (Pirnie *et al.*, 2003). Commonly used biodosimeters are spores of *B. subtilis* or MS2 phages. Cabaj *et al.* (1996), however, clearly demonstrated the dependency of the fluence distribution in a CF-system, the REF-value and the inactivation rate constant of the used model organism. REF decreases with increased fluence distribution and susceptibility (k-value) of the used model organism. MS2 phages and spores of *B. subtilis* are less sensitive to UV than most other pathogenic micro-organisms (Tables 7.1, 7.2, 7.3 and 7.4). The EPA manual (Pirnie *et al.*, 2003) introduced a REF bias based on effects of fluence distribution and inactivation rate constants to account for the difference in sensitivity between model organism and target pathogens. Another approach is the use of alternative model organisms. *E. coli* is suggested and also this review indicates that it can be used as model for the more susceptible bacteria and also for *Cryptosporidium* and *Giardia.* More recently Clancy *et al.* (2004) suggested two potential bacteriophages Qβ and T7 as model organisms. The use of T7 as UV dosimeter was previously proposed by Rontó *et al.* (1992). The k-values of these organisms (Table 7.2) are more in the range of the k-values calculated for the more sensitive pathogens.

7.1.6 Conclusions and further research

The accumulated literature data on the inactivation kinetics of disinfection with UV irradiation demonstrate that the process is effective against all pathogenic micro-organisms relevant for the current drinking water practices. The inactivation of micro-organisms by UV could be described with first-order kinetics using fluence-inactivation data from laboratory studies in collimated beam tests. No inactivation at low fluences (shoulder) and reduced inactivation at higher fluences (tailing) was observed to some extent. The former deviation from the log-linear kinetics is included in MIC calculations and the latter affects the maximum MIC-values. The parameters that were used to describe the inactivation are the inactivation rate constant k (cm^2/mJ), the maximum inactivation demonstrated and (only for bacterial spores and Acanthamoeba) the offset parameter. The most persistent organisms known are viruses and specifically Adenoviruses and bacterial spores. From the protozoa Acanthamoeba was highly UV resistant. Bacteria and (oo)cysts of Cryptosporidium and Giardia are more susceptible with fluence requirement of <20 mJ/cm^2 for a MIC of 3 log.

Research items. Based on this review a number of knowledge gaps are identified. More quantitative information is needed to estimate the effect of micro-organism related factors like environmental species and DNA-repair influencing the fluence requirement of UV disinfection under full-scale conditions. Biodosimetry is a powerful tool to determine germicidal fluence values of CF-systems, but to determine REF for the whole range of relevant micro-organisms with different UV sensitivities, additional model organisms are needed. In the application of medium-pressure lamps with polychromatic UV light, fluence assessment has to be developed further using biodosimetry in connection with fluence calculation models. Daily UV process control needs further research to obtain simple, reliable and cheap *insitu* process parameters to monitor the efficacy of UV systems in practice.

8 Overall MEC-matrix for drinking water processes and further research

From the results of the individual evaluated processes a total overview of Micro-organisms Elimination Credit (MEC)-values is made for physical processes in order to compare the efficacy of the different water treatment processes used for drinking water production. The range of observed Decimal Elimination Capacity (DEC)-values used for calculating the MEC-value shows the variability of MEC. The MEC-value is qualified by showing the average full-scale (FS)-index and the product of FS with the number of data (Q).

For disinfection processes a required dose or fluence Table for a MIC of 1, 2, 3 and 4 log is created, based on the calculated inactivation kinetics and corrected for the effects of microbial or dose related factors affecting the efficacy of the inactivation under full-scale conditions.

8.1 Physical processes

MEC-values. In Table 8.1 the MEC-values of conventional treatment, coagulation/floc-removal, rapid granular filtration and slow sand filtration are presented for viruses, bacteria, bacterial spores and *Cryptosporidium* and *Giardia*. The range in DEC observed in the studies was considerable (Figure 8.1) and is caused by several process conditions described in the separate chapters.

Based on theses data MEC of conventional treatment (coagulation and filtration) and slow sand filtration for viruses, bacteria and bacterial spores is in the same order of magnitude (2–3 log), which is approximately 1 log less as calculated for both parasitic protozoa. Furthermore, MEC of the combination of coagulation/filtration by in line coagulation (RGFicoa) for viruses and bacteria is lower than for conventional treatment (combination of separate processes). For the spores and parasitic protozoa, however, MEC of RGFicoa and conventional treatment was in the same order of magnitude. The effectiveness of coagulation/floc-removal for the removal of the five groups of micro-organisms was similar (around 1.5 log). RGF operated as part of a conventional treatment (RGFcoa) removes viruses and bacteria with 1 log, approximately 0.5 log less than spores and protozoa. Of the physical processes RGF operated without a preceding coagulation or filled with granular activated carbon showed the lowest MEC-values of 0.5–1.0 log for the different groups of pathogenic micro-organisms.

Qualification of the MEC values and further research. The FS-index values show that for bacterial spores experimental conditions approach the full-scale conditions (FS = 3.8–5.0). This is mainly because in the weighing factor the surrogate weighing does not count for this group of organisms. For the other groups of micro-organisms conditions were moderately comparable to the full-scale conditions (FS = 3.0).

Table 8.1 Micro-organism elimination credit (MEC), the range of DEC-values of treatment processes used in drinking water production and the quality of the database (Q and FS)

	Data	Conventional treatment	Coagulation/ floc-removal	RGF	RGFcoa	RGFicoa	RGFgac	Slow sand filtration
Viruses	**MEC**	**3.0**	**1.8**	**0.8**	**1.1**	**0.9**	**0.5**	**2.2**
	Range	1.2–5.3	0.2–4.3	0.1–3.8	0.2–2.5	0.1–1.5	0.2–0.7	0.6–4.0
	FS	3.6	3.5	3.0	3.1	3.0	2.5	3.3
	Q	248	**311**	270	102	270	24	43
Bacteria	**MEC**	**2.1**	**1.5**	**0.5**	**0.9**	**1.4**	**0.8**	**2.7**
	Range	1.0–3.4	0.6–3.7	0.2–1.0	0.4–1.5	0.8–2.1	0.7–0.9	1.2–4.8
	FS	3.1	3.0	2.8	3.0	3.0	3.0	3.4
	Q	167	**303**	**1268**	180	93	45	58
Bacterial spores	**MEC**	**2.4**	**1.4**	**0.7**	**1.6**	**2.3**	**0.7**	**1.5**
	Range	1.4–4.7	0.8–3.2	0.0–1.4	0.5–2.9	1.5–3.9	0.1–1.5	0.0–4.0
	FS	4.7	4.6	4.1	4.1	3.8	5.0	4.4
	Q	291	**423**	**435**	**504**	114	**420**	40
Cryptosporidium	**MEC**	**3.2**	**1.9**	**1.1**	**1.1**	**3.0**	**0.9**	**4.8**
	Range	1.4–5.5	0.4–3.8	0.0–2.3	0.8–5.5	0.1–5.4	0.7–1.1	2.7–6.5
	FS	3.7	3.3	3.1	3.5	2.7	3.0	3.4
	Q	**599**	**304**	143	**364**	**643**	26	27
Giardia	**MEC**	**3.4**	**1.6**	**1.2**	**1.9**	**2.9**	**0.8**	**4.9**
	Range	2.1–5.1	0.3–2.9	0.0–2.6	0.3–4.1	0.8–4.7	0.4–1.3	4.0–6.0
	FS	4.3	3.9	3.5	3.2	3.0	3.0	3.7
	Q	288	207	60	**307**	**315**	10	11

Q-values presented in Table 8.1 most data were found for removal of *Cryptosporidium* by conventional treatment or RGFicoa. The product of FS-index and the number of data Q was 599 and 643, respectively. This is not surprising since the health risks of the parasitic protozoa, *Cryptosporidium* in particular, was a major cause for the increasing interest in micro-organism removal by water treatment. Generally, the MEC of these processes for parasitic protozoa is based on a sound set of data.

MEC-values of the coagulation process for the different organisms is in the same order of magnitude and based on a large amount of data with Q-values of >207. The MEC-values for the rapid granular filtration without a preceding coagulation for viruses, bacteria and bacterial spores was based on a high qualified dataset (Q 270–1268) but for both parasitic protozoa the datasets were limited (Q 60–143). For RGF with preceding coagulation except for bacterial spores this was the other way around. Datasets for parasites and viruses for RGF with in line coagulation (RGFicoa) had a high Q-value (270 -643) but for bacteria and bacterial spores the datasets were limited (93–114). Finally, except for bacterial spore removal by RGFgac the datasets for RGFgac and SSF for all micro-organisms is qualified as limited with Q varying between 10 and 58.

Process conditions:

- The strategy to control DEC of a local conventional treatment is developed and mainly based on the control of breakthrough in a RGF during a ripening period (after back washing) and during the end of the operational period with increasing head loss. To define the operational measures to control breakthrough, surrogates are available like turbidity, particle counting, bacterial spores and algal cells. The first three of these surrogates are too conservative for assessing DEC of RGF; a correction factor of 0.5–1.0 log must be applied. When the filter bed operation is optimised for protozoa removal this will benefit also the minimization of the breakthrough of bacteria and viruses, although it is expected that for the latter breakthrough is of less influence.

– Slow sand filtration can be regarded as a robust process with over time a relative low variability in DEC. On the basis of a workshop in November 2006 (Hijnen *et al.*, 2006) the dataset will be extended with data on the influence of temperature and biological processes in the filter bed (i.e. predation), the filtration rate (economically driven) and alternative monitoring strategies to control the DEC in practice.

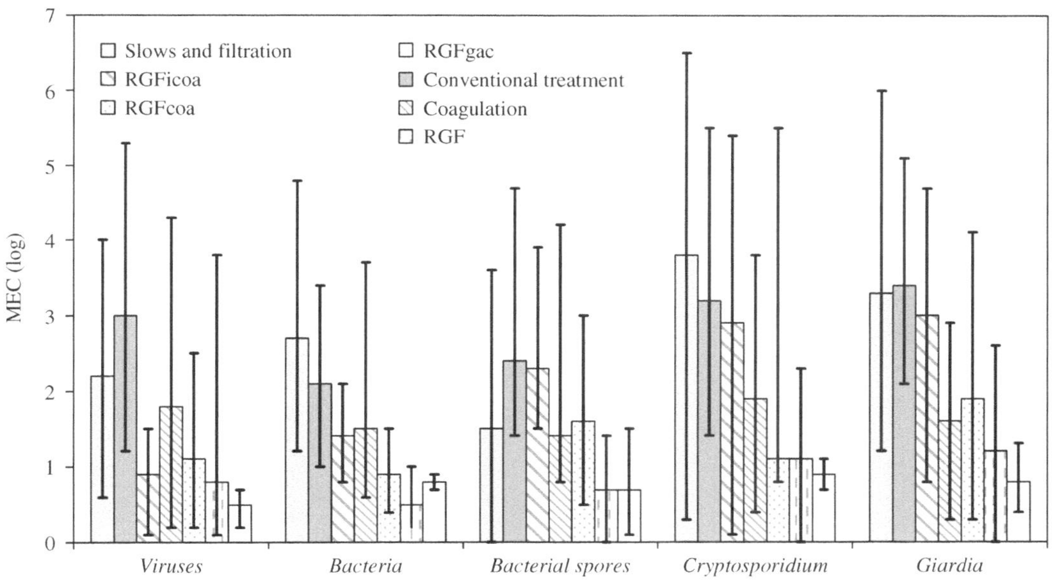

Figure 8.1 The MEC-values of treatment processes for the different micro-organisms (error bars represent the variation in DEC)

8.2 Disinfection processes

8.2.1 UV-disinfection: required UV-fluence

The accumulated knowledge in this review is used to create the required fluence (MIC) 1, 2, 3 or 4 log inactivation for most micro-organisms relevant for microbiological safety of water (Table 8.2). The required fluence is calculated based on the k-values presented in Tables 7.1, 7.3 and 7.4 with correction for the increased UV-sensitivity of environmental bacteria and bacterial spores. For five species this was based on specific literature data and for the other bacteria where no data were available, this was set at a factor of three. This correction was not necessary for the viruses (see data on FRNA-phages in Figure 7.7) and for the protozoa no data are available for such a correction. Increased fluence requirement because of DNA-repair did not seem necessary for viruses and protozoa. For bacteria dark repair, relevant for drinking water disinfection, will not have a significant effect on the required fluence, but due to photo-reactivation in wastewater disinfection, fluence requirement for this application will increase with a factor of 4.

Table 8.2 The UV fluence (mJ/cm^2) requirements for a MIC of 1 up to 4 log by monochromatic UV radiation for viruses, bacteria, bacterial spores and protozoan (oo)cysts based on the k-values with or without correction for environmental species; for bacteria in wastewater an extra correction of a factor of 4 is required for photo-reactivation; for Giardia increased fluence requirement because of dark repair is a factor for further research

	UV fluence (mJ/cm^2) requirements			
MIC (log) required	1	2	3	4
Bacillus subtilis[a]	56	111	167	222
Adenovirus type 40	56	111	167	–[d]
Clostridium perfringens[a]	45	95	145	–[d]
Adenovirus type 2,15,40,41	42	83	125	167
Acanthamoeba[c]	40	71	119	167
Adenovirus[a] (no type 40)	25	50	–[d]	–[d]
Calicivirus canine	10	21	31	41
Rotavirus SA-11	10	20	29	39
Calicivirus feline	9	19	28	38
Coxsackie virus B5	8	17	25	34
Streptococcus faecalis[a]	9	16	23	30
Legionella pneumophila[b]	8	15	23	30
Poliovirus type 2	7	15	22	30
Shigella sonnei[b]	6	13	19	26
Salmonella typhi[a]	6	12	17	51
Hepatitis A	6	11	17	22
Calicivirus bovine	5	11	16	21
E. coli O157[b]	5	9	14	19
E. coli[a]	5	9	14	18
Cryptosporidium EPA[c]	3	6	12	–[e]
Giardia EPA[c]	2	5	11	–[e]
Campylobacter jejuni[b]	3	7	10	14
Yersinia enterocolitica[b]	3	7	10	13
Legionella pneumophila[b]	3	6	8	11
Shigella dysenteriae [b]	3	5	8	11
Vibrio cholerae[b]	2	4	7	9

[a]environmental spp.; [b]corrected for environmental spp.; [c]no correction for environmental spp. (research needed); [d]MIC_{max} <4 log; [e]no value due to tailing

9 Literature

9.1 Introduction and data collection and evaluation methods

Gale, P., Pitchers, R., and Gray, P. (2002). The effect of drinking water treatment on the spatial heterogeneity of micro-organisms: implications for assessment of treatment efficiency and health risk. *Wat. Res.* **36**, 1640–1648.

Hijnen, W.A.M, Veenendaal, D., van der Speld, W.M.H., Visser, A., Hoogenboezem, W., and van der Kooij, D. (2000). Enumeration of faecal indicator bacteria in large water volumes using on site membrane filtration to assess water treatment efficiency. *Wat. Res.* **34**(5), 1659–1665.

Hijnen, W.A.M., Medema, G.J., and van der Kooij, D. (2004). Quantitative assessment of the removal of indicator-bacteria by full-scale treatment. *Wat. Sci. Technol.* **4**(2), 47–54.

LeChevallier, M.W., Norton, W.D., and Lee, R.G. (1991). Occurrence of *Giardia* and *Cryptosporidium* spp. In surface water supplies. *Appl. Environ. Microbiol.* **57**(9), 2610–2616.

Medema, G.J., Hoogenboezem, W., van der Veer, A.J., Ketelaars, H.A.M., Hijnen, W.A.M., and Nobel, P.J. (2003). Quantitative risk assessment of *Cryptosporidium* in surface water treatment. *Wat. Sci. Technol.* **47**(3), 241–247.

Payment, P., Trudel, M., and Plante, R. (1985). Elimination of viruses and indicator bacteria at each step of treatment during preparation of drinking water at seven water treatment plants. *App. Environ. Microbiol.* **49**(6), 1418–1428.

Payment, P. and Franco, E. (1993). *Clostridium perfringens* and somatic coliphages as indicators of the efficiency of drinking water treatment for viruses and protozoan cysts. *Appl. Environ. Microbiol.* **59**(8), 2418–2424.

9.2 Data collection and evaluation methods

Chick, H. (1908). An investigation of the laws of disinfection. *J. Hyg.*, **8**, 92–157.

Hoyer, O. (1998). Testing performance and monitoring of UV systems for drinking water disinfection. *Wat. Supply* **16**(1/2), 424–429.

Hoyer, O. (2004). Water disinfection with UV radiation – requirements and realization. In *Proceedings European Conference UV Karlsruhe, UV radiation. Effects and Technologies*, September 22–24, 2003, Karlsruhe

Knudson, G.B. (1985). Photoreactivation of UV irradiated *Legionella Pneumophila* and other *Legionella* species. *Appl. Environ. Microbiol.* **49**(4), 975–980.

Mamane-Gravetz, H. and Linden, K.G. (2005). Relationship between physiochemical properties, aggregation and UV inactivation of isolated environmental spores in water. *J. Appl. Microbiol.* **98**, 351–363.

Österreichisches Normungsinstitut (1999). Anlagen zur Desinfektion von Wasser mittels Ultraviolett-strahlen, Anforderungen und Prüfung. Önorm M 5873-1, Wien.

Pirnie, M., Linden, K.G., and Malley, J.P. (2003). Ultraviolet disinfection guidance manual. EPA 815-D-03-007, June 2003.

Qualls, R.G. and Johnson, J.D. (1983a). Bioassay and fluence measurement in UV disinfection. *Appl. Environ. Microbiol.* **45**, 872–877.

Severin, B.F., Suidan, M.T., and Engelbrecht, R.S. (1983). Effects of temperature on ultraviolet light disinfection. *Environ. Sci. Technol.* **17**, 717–721.

Sommer, R., Haider, Th., Cabaj, A., Pribil, W., and Lhotsky, M. (1998). Time fluence reciprocity in UV disinfection of water. *Wat. Sci. Tech.* **38**(12), 145–150.

Sommer, R., Cabaj, A., Sandu, T., and Lhotsky, M. (1999). Measurement of UV radiation using suspensions of micro-organisms. *J. Photochem. Photobiol. B:Biol.* **53**, 1–6.

von Sonntag, C., Kolch, A., Gebel, J., Oguma, K., and Sommer, R. (2004). The photochemical basis of UV disinfection. In *Proc. European Conference UV Karlsruhe, UV radiation. Effects and Technologies*, September 22–24, 2003, Karlsruhe.

Watson, H.E. (1908). A Note on the variation of the rate of disinfection with change in the concentration of the disinfectant. *J. Hyg.* **8**, 536–542.

9.3 Conventional treatment

Akiba, M., Kunikane, S., Kim, H.S., and Kitazawa, H. (2002). Algae as surrogate indices for the removal of Cryptosporidium oocysts by direct filtration. *Wat. Sci. Tech.:Water Suppl.* **2**(30), 73–80.

Cornwell, D.A. and Macphee, M.J. (2001). Effects of spent filter backwash recycle on Cryptosporidium removal. *Jour. AWWA* April, 2001, 153–162.

Dugan, N.R., Fox, K.R., Owens, J.H., and Miltner, R.J. (2001). Controlling *Cryptosporidium* Oocysts using conventional treatment. *Journal of the AWWA* **93**(12), 64–76.

Dugan, N.R. and Williams, D.J. (2004). Removal of Cryptosporidium by in-line filtration: effects of coagulant type, filter loading rate and temperature. *Jour. Wat. Suppl.: Aqua* **53**(1), 1–15.

Foliguet, J.M. and Doncoeur, F. (1975). Elimination des enteroviruses au cours du traitement des eaux d'alimentation par coagulation-flocculation-filtration. *Wat. Res.* **9**, 953–961.

Guy, M.D., McIver, J.D., and Lewis, M.J. (1977). The removal of virus by a pilot plant treatment plant. *Wat. Res.* **11**, 421–428.

Hancock, C.M., Ward, J.V., Hancock, K.W., Klonocki, P.T., and Sturbaum, G.D. (1996). Assessing plant performance using MPA. *JAWWA* **88**(12), 24–34.

Hashimoto, A., Hirata, T., and Kunikane, S. (2001). Occurrence of Cryptosporidium oocysts and Giardia cysts in a conventional water purification plant. *Wat. Sci. Techn.* **43**(12), 89–92.

Harrington, G.W., Xagoraraki, I., Assavasilavasukul, P., and Standridge, J.H. (2003). Effect of filtration conditions on removal of emerging waterborne pathogens. *Journ. AWWA* **95**(12), 95–104.

Havelaar, A.H., van Olphen, M., and Schijven, J.F. (1995). Removal and inactivation of viruses by drinking water treatment processes under full scale conditions. *Wat. Sci. Tech.* **31**(5–6), 55–62.

Hijnen, W.A.M., van der Kooij, D., Drost, Y.C., and Veenendaal, H., and van der Leer, R.Chr. (1994). Removal of micro-organisms by drinking water treatment at production location Zevenbergen. Kiwa report, SWO 94.254, in Dutch.

Hijnen, W.A.M., van der Speld, W.M.H., Houtepen, F.A.P., and van der Kooij, D. (1997). Spores of sulfite-reducing clostridia: a surrogate parameter for assessing the effects of water treatment on protozoan oocysts? In: *1997 International Symposium on waterborne Cryptosporidium*, C.R. Fricker, J.L. Clancy and P.A. Rochelle (eds.), American Water Works Association, Denver USA, pp. 115–126.

Hijnen, W.A.M., Hoogenboezem, W., van der Kooij, D., and Medema, G.J. (1998a). Removal of indicator bacteria during drinking water production at production location Andijk. Kiwa report SWI 98.185, in Dutch.

Hijnen, W.A.M., Nobel, P.J., Westra, R., van der Kooij, D., and Medema, G.J. (1998b). Removal of indicator bacteria during drinking water production at three locations of Water Company Europoort. Kiwa report 98.160, in Dutch.

Hijnen, W.A.M., van der Kooij, D., and Medema, G.J. (1998c). Removal of indicator bacteria during drinking water production at location Braakman (DELTA). Kiwa report, 98.177.

Hijnen, W.A.M., Vrouwenvelder, J.S., and Manolarakis, S.A. (2003). Microbiological aspects of the supply of House Hold Water. Kiwa report KWR 02.095B, in Dutch.

Jekel, M.T. (1994). Flocculation effects of ozone. *Ozone Sci. and Eng.* **16,** 55–66.

LeChevallier, M.W. and Norton, W.D. (1992). Examining relationships between particle counts, and Giardia and Cryptosporidium and turbidity. *J. Am. Water Works Assoc.* **84**(12), 54B60 (1992).

Logsdon, G.S., Symons, J.M., Hoye, Jr. R.L., and Arozarena, M.M. (1981). Alternative filtration methods for removal of Giardia cysts and cyst models. *Jour. AWWA* **1981**(2), 111.

Nieminski, E.C. (1994). Giardia and Cryptosporidium removal through direct filtration and conventional treatment. In *Proc. AWWA Annual Conference*, 1994.

Nieminski, E.C. and Ongerth, J.E. (1995). Removing Giardia and Cryptosporidium by conventional treatment and direct filtration. *Jour. AWWA* **87**(9), 96–106

Nobel, P.J. and Medema, G.J. (1999). Removal of surrogates in treatment: pilot plant WBE. Kiwa report SWI 99.180, in Dutch.

Payment, P., Trudel, M., and Plante, R. (1985). Elimination of viruses and indicator bacteria at each step of treatment during preparation of drinking water at seven water treatment plants. *App. Environ. Microbiol.* **49**(6), 1418–1428.

Payment, P. and Franco, E. (1993). *Clostridium perfringens* and somatic coliphages as indicators of the efficiency of drinking water treatment for viruses and protozoan cysts. *Appl. Environ. Microbiol.* **59**(8), 2418–2424.

Patania, N.L., Jacangelo, J.G., Cummings, L., Wilczak, A., Riley, K., and Oppenheimer, J. (1995). Optimization of Filtration for Cyst Removal. Denver, Colo: Awwa Research Foundation and American Water Works Association.

Rice, E.W., Fox, K.R., Miltner, R.J., Lytle, D.A., and Johnson, C.H. (1996). Evaluating plant performance with endospores. *JAWWA* **88**(9), 122–130.

Saunier, B.M., Selleck, R.E., and Trussell, R.R. (1983). Preozonation as a coagulant aid in drinking water treatment. *JAWWA* **75**(5), 239–246.

States, S., Stadterman, K., Ammon, L., Vogel, P., Baldizar, J., Wright, D., Conley, L., and Sykora, J. (1997). Protozoa in river water: sources, occurrence and treatment. *JAWWA* **89**(9), 74–83.

West, Th., Daniel, P., Meyerhofer, P., DeGraca, A., Leonard, S., and Gerba, C. (1994). Evaluation of Cryptosporidium removal through high-rate filtration. In *Proc. Annual Conf. AWWA*, Denver CO.

Yates, R.S., Green, J.F., Liang, S., Merlo, R.P., and De Leon, R. (1997). Optimizing coagulation/filtration processes for Cryptosporidium removal. *Int. Symp. on Waterborne Cryptosporidium*, C.R. Fricker, J.J. Clancy and P.A. Rochelle (eds), Newport Beach, US, March 1997, pp. 281–290.

9.4 Coagulation flocculation

Akiba, M., Kunikane, S., Kim, H.S., and Kitazawa, H. (2002). Algae as surrogate indices for the removal of Cryptosporidium oocysts by direct filtration. *Wat. Sci. Tech.: Water Suppl.* **2**(30), 73–80.

Bell, 2000 in LeChevallier 2001 WHO-document.

Cornwell, D.A. and Macphee, M.J. (2001). Effects of spent filter backwash recycle on Cryptosporidium removal. *Jour. AWWA* 153–162.

Dugan, N.R., Fox, K.R., Owens, J.H., and Miltner, R.J. (2001). Controlling *Cryptosporidium* Oocysts using conventional treatment. *Journal of the AWWA* **93**(12), 64–76.

Edzwald, J.K. and Kelley, M.B. (1998). Control of Cryptosporidium: from reservoirs to clarifiers to filters. *Wat. Sci. Tech.* **37**(20), 1–8.

Edzwald, J.K., Tobiason, J.E., Dunn, H., Kaminski, G., and Galant, P. (2001). Removal and fate of *Cryptosporidium* in dissolved air drinking water treatment plants. *Wat. Sci. and Techn.* **43**(8), 51–57.

Edzwald, J.K. and Tobiason, J.E. (2002). Fate and removal of *Cryptosporidium* in a dissolved air flotation water plant with and without recycle of waste filter back wash water. *Wat. Sci. and Technol.: Wat. Supply* **2**(2), 85–90.

Edzwald, J.K., Tobiason, J.E., Udden, C.T., Kaminski, G.S., Dunn, H.J., Galant, P.B., and Kelley, M.B. (2003). Evaluation of the effect of recycle of waste filter backwash water on plant removals of Cryptosporidium. *Jour. Water Supply: Aqua*, **52**(4), 243–258.

Gray, P., Pitchers, R., Stanfield, G., Camm, R., Connor, K., and Hall, T. (1997). Quantifying the removal of microbes during water treatment. WRc, report PT 2073.

Haas, C.N., French, K., Finch, G.R., and Guest, R.K. (2001). Data review on the physical/chemical removal of *Cryptosporidium*. Denver, Colo.: Am. Wat. Works Assoc. Res. Foundation.

Havelaar, A.H., van Olphen, M., and Schijven, J.F. (1995). Removal and inactivation of viruses by drinking water treatment processes under full scale conditions. *Wat. Sci. Tech.* **31**(5–6), 55–62.

Hijnen, W.A.M., van der Kooij, D., Drost, Y.C., Veenendaal, H., and van der Leer, R.Chr. (1994). Removal of micro-organisms by drinking water treatment at production location Zevenbergen. Kiwa report, SWO 94.254, in Dutch.

Hijnen, W.A.M., van der Speld, W.M.H., Houtepen, F.A.P., and van der Kooij, D. (1997). Spores of sulfite-reducing clostridia: a surrogate parameter for assessing the effects of water treatment on protozoan oocysts? In: *1997* International Symposium on waterborne Cryptosporidium, C.R. Fricker, J.L. Clancy and P.A. Rochelle (eds), American Water Works Association, Denver USA, pp. 115–126.

Hijnen, W.A.M., Nobel, P.J., Westra, R., van der Kooij, D., and Medema, G.J. (1998b). Removal of indicator bacteria during drinking water production at three locations of Water Company Europoort. Kiwa report 98.160, in Dutch.

Hijnen, W.A.M., van der Kooij, D., and Medema, G.J. (1998c). Removal of indicator bacteria during drinking water production at location Braakman (DELTA). Kiwa report, 98.177.

Hijnen, W.A.M., Wübbels, G.H., Soppe, A.I.A., Wolters, H., Nobel, P.N., van der Kooij, D., and Medema, G.J. (1998d). Removal of indicator bacteria during drinking water production at location De Punt, Water Company Groningen. Kiwa report 98.178.

Hijnen, W.A.M., Medema, G.J., and van der Kooij, D. (2002). Quantitative assessment of the removal of indicator-bacteria by full-scale treatment. In: *Proc. Int. Symp. on Waterborne Pathogens*, Lissabon, September 2002.

Jekel, M.T. (1994). Flocculation effects of ozone. *Ozone Sci. and Eng.* **16,** 55–66.

Kelley, M.B., Warrier, P.W., Brokaw, J.K., Barrett, K.L., and Komisar, S.J. (1995). A study of two I.S. Army installation drinking water sources and treatment systems for the removal of *Giardia* and *Cryptosporidium.* Water Quality Technology Conference, November 12–15, New Orleans, LA, US.

Kelley, M.B. (1996). The removal of *Cryptosporidium* by selected drinking water treatment processes. Thesis, Rensselaer Polytechnic Institute, Troy, New York.

LeChevallier, M.W., Norton, W.D., Lee, R.G., and Rose, J.B. (1991). *Giardia* and *Cryptosporidium* in water Supplies. Denver, Colo: Am. Wat. Works Assoc. Res. Foundation.

LeChevallier, M.W. (2001). Impact of treatment on microbial water quality: a review document on treatment efficiency to remove pathogens. WHO document.

Logsdon, G.S., Thurman, V.C., Frindt, E.S., and Stoecker, J.G. (1985). Evaluating sedimentation and various filter media for removal of Giardia cysts. *J. AWWA* 61–66.

Nasser, A., Weinberg, D., Dinoor, N., Fattal, B., and Adin, A. (1995). Removal of Hepatitis A virus (HAV), Poliovirus and MS2 coliphage by coagulation and high rate filtration. *Wat. Sci. Tech.* **31**(5–6), 63–68.

Patania, N.L., Jacangelo, J.G., L. Cummings, L., Wilczak, A., Riley, K., and Oppenheimer, J. (1995). Optimization of Filtration for Cyst Removal. Denver, Colo: Awwa Research Foundation and American Water Works Association.

Payment, P., Trudel, M., and Plante, R. (1985). Elimination of viruses and indicator bacteria at each step of treatment during preparation of drinking water at seven water treatment plants. *App. Environ. Microbiol.* **49**(6), 1418–1428.

Payment, P. and Franco E. (1993). *Clostridium perfringens* and somatic coliphages as indicators of the efficiency of drinking water treatment for viruses and protozoan cysts. *Appl. Environ. Microbiol.* **59**(8), 2418–2424.

Plummer, J.D., Edzwald, J.K., and Kelley, M.B. (1995). Removing Cryptosporidium by dissolved air flotation. *Jour. AWWA* **87**(9), 85–95.

Rao, V.Ch., Symons, J.M., Ling, A., Wang, P., Metcalf, Th.G., Hoff, J.C., and Melnick, J.L. (1988). Removal of Hepatitis A virus and rotavirus by drinking water treatment. *Jour. AWWA* 59–67.

Rice, E.W., Fox, K.R., Miltner, R.J., Lytle, D.A., and Johnson, C.H. (1996). Evaluating plant performance with endospores. *JAWWA* **88**(9), 122–130.

Saunier, B.M., Selleck, R.E., and Trussell, R.R. (1983). Preozonation as a coagulant aid in drinking water treatment. *JAWWA* **75**(5), 239–246.

States, S., Stadterman, K., Ammon, L., Vogel, P., Baldizar, J., Wright, D., Conley, L., and Sykora, J. (1997). Protozoa in river water: sources, occurrence and treatment. *Jour. AWWA* **89**(9), 74–83.

Stetler, R.E., Waltrip, S.C., and Hurst, C.J. (1992). Virus removal and recovery in the drinking water treatment train. *Wat. Res.* **26**(6), 727–731.

van Puffelen, J., Buijs, P.J., Nuhn, P.V.A.M., and Hijnen, W.A.M. (1995). Dissolved air flotation in potable water treatment: the Dutch experience. *Wat. Sci. Tech.* **31**(3–4), 149–157.

Wiersema, I. (1997). Influence of process factors on the elimination of indicator bacteria by coagulation processes used in drinking water production. Report 97–35, Wageningen University, The Netherlands, in Dutch.

9.5 Rapid granular filtration

Al-Ani, M., McElroy, J.M., Hibler, Ch.P., and Hendricks, D.W. (1985). Filtration of Giardia cysts and other substances: Volume 3. Rapid-rate filtration. EPS Project summary.

Akiba, M., Kunikane, S., Kim, H.S., and Kitazawa, H. (2002). Algae as surrogate indices for the removal of *Cryptosporidium* oocysts by direct filtration. *Wat. Sci. Tech.:Water suppl.* **2**(30), 73–80.

Chung, J., Hijnen, W.A.M., Vesey, G., and Ashbolt, N.J. (2004). Potential Cryptosporidiun oocysts surrogates for sand filtration and the importance of their surface properties. International *Giardia* and *Cryptosporidium* Conference, Amsterdam, The Netherlands.

Cleasby, J.L., Hilmoe, D.J., and Dimitracopoulos, C.J. (1984). Slow sand and direct in-line filtration of a surface *waster. Jour. AWWA* **76**(12), 44–55.

Coffey, B.M., Huck, P.M., Maurizo, D.D., Emelko, M.B., Douglas, I.P., and Van den Oever, J. (1999). The effect of optimizing coagulation on the removal of *Cryptosporidium parvum* and *Bacillus subtilis*. In *Proc. AWWA Water Quality Technology Conference*, Tampa, Florida.

Cornwell, D.A. and Macphee, M.J. (2001). Effects of spent filter backwash recycle on *Cryptosporidium* removal. *Jour. AWWA* 153–162.

Dai, X. and Hozalski, R.M. (2002). Effect of NOM and biofilm on the removal of *Cryptosporidium parvum* oocysts in rapid filters. *Wat. Res.* **36**, 3523–3532.

Denny, S. and Broberg, P. (1992). Microbiological efficiency of water treatment: microbiological monitoring of full-scale water treatment processes. Foundation for Water Research, report FR0282.

De Leon, R., Singh, S.N., Rose, J.B., Mullinax, L., Musial, C.E., Kutz, S.M., Sinclair, N.A., and Gerba, C.P. (1986). Micro-organisms removal from wastewater by rapid mixed media filtration. *Wat. Res.* **20**(5), 583–587.

Dugan, N., Fox, K., Owens, J., and Miltner, R. (2001). Controlling *Cryptosporidium* Oocysts using conventional treatment. *Journal of the AWWA* **93**(12), 64–76.

Dugan, N.R. and Williams, D.J. (2004). Removal of *Cryptosporidium* by in-line filtration: effects of coagulant type, filter loading rate and temperature. *Jour. Wat. Suppl.: Aqua* **53**(1), 1–15.

Dugan, N.R. and Williams, D.J. (2004). Removal of *Cryptosporidium* by in-line filtration: effects of coagulant type, filter loading rate and temperature. *Jour. Wat. Suppl.: Aqua* **53**(10), 1–8.

Emelko, M.B. (2001). Removal of *Cryptosporidium parvum* by granular media filtration. Thesis, University of Waterloo, Ontario, Canada, 2001.

Emelko, M.B. (2003). Removal of viable and inactivated *Cryptosporidium* by dual- and tri-media filtration. *Wat. Res.* **37**, 2998–3008.

Foliguet, J.M. and Doncoeur, F. (1975). Elimination des enteroviruses au cours du traitement des eaux d' alimentation par coagulation-flocculation-filtration. *Wat. Res.* **9**, 953–961.

Gitis, V., Adin, A., Nasser, A., Gun, J., and Lev, O. (2002). Fluorescent dye labelled bacteriophages- a new tracer for the investigation of viral transport in porous media: 2. studies of deep-bed filtration. *Wat. Res.* **36**, 4235–4242.

Gitis, V., Haught, R.C., Clark, R.M., and Krishnan, E.R. (2002). Depressed filtration ripening enhances removal of *Cryptosporidium parvum. Wat. Sci. Technol.: Wat. Suppl.* **2**(30), 159–168.

Gray, P., Pitchers, R., Stanfield, G., Camm, R., Connor, K., and Hall, T. (1997). Quantifying the removal of microbes during water treatment. WRc, report PT 2073.

Guy, M.D., McIver, J.D., and Lewis, M.J. (1977). Removal of virus by a pilot treatment plant. *Wat. Res.* **11**, 421–428.

Haas, Ch.N., Severin, B.F., Roy, D., Engelbrecht, R.S., and Lalchandani, A. (1985). Removal of new indicators by coagulation and filtration. *Jour. AWWA* (2) 1985.

Hall, T., Pressdee, J., Gregory, R., and Murray, K. (1995). *Cryptosporidium* removal during water treatment using dissolved air flotation. *Wat. Sci. Tech.* **31**(3–4), 125.

Hancock, C.M., Ward, J.V., Hancock, K.W., Klonocki, P.T., and Sturbaum, G.D. (1996). Assessing plant performance using MPA. *JAWWA* **88**(12), 24–34.

Hashimoto, A., Hirata, T., and Kunikane, S. (2001). Occurrence of *Cryptosporidium* oocysts and *Giardia* cysts in a conventional water purification plant. *Wat. Sci. Techn.* **43**(12), 89–92.

Havelaar, A.H., van Olphen, M., and Schijven, J.F. (1995). Removal and inactivation of viruses by drinking water treatment processes under full scale conditions. *Wat. Sci. Tech.* **31**(5–6), 55–62.

Hijnen, W.A.M., van der Kooij, D., Drost, Y.C., Veenendaal, H., and van der Leer, R.Chr. (1994). Removal of micro-organisms by drinking water treatment at production location Zevenbergen. Kiwa report, SWO 94.254, in Dutch.

Hijnen, W.A.M., Koolen, R., van der Speld, W.M.H., and Kreukniet, A.A. (1997a). Removal of microbial and physical parameters by a rapid sand filter as surrogates for pathogenic micro-organisms. Concept Kiwa report, in Dutch.

Hijnen, W.A.M., van der Speld, W.M.H., Houtepen, F.A.P., and van der Kooij, D. (1997b). Spores of sulfite-reducing clostridia: a surrogate parameter for assessing the effects of water treatment on protozoan oocysts? In: *1997 International Symposium on Waterborne Cryptosporidium*, C.R. Fricker, J.L. Clancy and P.A. Rochelle (eds), American Water Works Association, Denver USA, pp. 115–126.

Hijnen, W.A.M., Hoogenboezem, W., van der Kooij, D., and Medema, G.J. (1998a). Removal of indicator bacteria during drinking water production at production location Andijk. Kiwa report SWI 98.185, in Dutch.

Hijnen, W.A.M., Magic, A., te Welscher, R.A.G., Krom, A., and Medema, G.J. (1998b). Removal of indicator bacteria during drinking water production at two locations of Amsterdam Water works. Kiwa report SWI 98.197, in Dutch.

Hijnen, W.A.M., Nobel, P.J., Westra, R., van der Kooij, D., and Medema, G.J. (1998c). Removal of indicator bacteria during drinking water production at three locations of Water Company Europoort. Kiwa report 98.160, in Dutch.

Hijnen, W.A.M., Visser, A., and Medema, G.J. (1998d). Removal of indicator bacteria during drinking water production at location Scheveningen, Water Company South-Holland. Kiwa report 98.216, in Dutch.

Hijnen, W.A.M., van der Kooij, D., and Medema, G.J. (1998e). Removal of indicator bacteria during drinking water production at location Braakman (DELTA). Kiwa report, 98.177.

Hijnen, W.A.M., Wübbels, G.H., Soppe, A.I.A., Wolters, H., Nobel, P.N., van der Kooij, D., and Medema, G.J. (1998f). Removal of indicator bacteria during drinking water production at location De Punt, Water Company Groningen. Kiwa report 98.178.

Hijnen, W.A.M., Meerkerk, M.A., and Buijs, P.J. (1999). Particle removal by treatment processes: experiences with particle counters and relation with micro-organisms. Kiwa report SWI 99.123, in Dutch.

Hijnen, W.A.M., Vrouwenvelder, J.S., and Manolarakis, S.A. (2003). Microbiological aspects of the supply of House Hold Water. Kiwa report KWR 02.095B, in Dutch.

Hijnen, W.A.M., Schijven, J.F., Bonné, P., Visser, A., and Medema, G.J. (2004). Elimination of viruses, bacteria and protozoan oocysts by slow sand filtration. *Wat. Sci. and Technol.* **50**(1), 147–154.

Hijnen, W.A.M., Brouwer-Hanzens, A.J., Charles, K.J., and Medema, G.J. (2005). Transport of MS2 phage, *Escherichia coli*, *Clostridium perfringens*, *Cryptosporidium parvum* and *Giardia intestinalis* in a gravel and a sandy soil. *Environ. Sci. & Technol.* **39**, 7860–7868.

Hijnen, W.A.M., and Medema, G.J. (2007). Microbial Elimination Capacity of conventional water treatment for viruses, bacteria and protozoan oocysts. WQTC 4–8 November, 2007, Charlotte, VS.

Hijnen, W.A.M., Suylen, G.M.H., Bahlman. J.A., Brouwer-Hanzens, A., and Medema, G.J. (2010). GAC adsoption filters as barriers for viruses, bacteria and protozoan (oo)cysts in water treatment. *Water Res.* **44**, 1224–1234.

Hsu, B., Huang, Ch., and Ruhsing Pan, J. (2001). Filtration behavior of *Giardia* and *Cryptosporidium*- ionic strength and pH effects. *Wat. Res.* **35**(16), 3777–3782.

Kelley, M.B., Warrier, P.K., Brokaw, J.K., Barrett, K.L., and Komisar, S.J. (1995). A study of two U.S. Army installation drinking water sources and treatment systems for the removal of Giardia and Cryptosporidium. In proc. of AWWA WQTC, 1995.

Kim, H.S., Kobayashi, Y., Akiba, M., and Kunikane, S. (2002). Evaluation of *Scenedesmus quadricauda* as a surrogate of *Cryptosporidium* oocyst removal in direct filtration. *Wat. Sci. Techn.: Wat. Suppl.* **2**(5–6), 395–402.

Koivunen, J., Siitonen, A., and Heinonen-Tanski, H. (2003). Elimination of enteric bacteria in biological-chemical wastewater treatment and tertiary filtration units. *Wat. Res.* **37**, 690–698.

LeChevallier, M.W. and Norton, W.D. (1992). Examining relationships between particle counts, and *Giardia* and *Cryptosporidium* and turbidity. *J. Am. Water Works Assoc.* **84**(12), 54B60 (1992).

Logsdon, G.S., Symons, J.M., Hoye, Jr. R.L., and Arozarena, M.M. (1981). Alternative filtration methods for removal of Giardia cysts and cyst models. *Jour. AWWA* (2), 111.

Logsdon, G.S., Thurman, V.C., Frindt, E.S., and Stoecker, J.G. (1985). Evaluating sedimentation and various filter media for removal of *Giardia* cysts. *J. AWWA* 61–66.

MacKenzie, W.R., Hoxie, N.J., Proctor, M.E., Gradus, S., Blair, K.A., Peterson, D.E., Kazmierczak, J.J., Addiss, D.G., Fox, K.R., Rose, J.B., and Davis, J.P. (1994). A massive outbreak in Milwaukee of Cryptosporidium infection transmitted through the public water supply. *New. Engl. J. Med.* **331**, 161–167.

Mazoua, S. and Chauveheid, E. (2005). Aerobic spore-forming bacteria for assessing quality of drinking water produced from surface water. *Water Res.* **39**(20), 5186–5198.

Nasser, A., Weinberg, D., Dinoor, N., Fattal. B., and Adin, A. (1995). Removal of Hepatitis A virus (HAV), Poliovirus and MS2 coliphage by coagulation and high rate filtration. *Wat. Sci. Tech.* **31**(5–6), 63–68.

Nestor, I. and Costin, L. (1971). The removal of Coxsackie virus from water by sand obtained from the rapid sand filters of water-plants. *Jour. Hyg. Epi. Microbiol. and Immun.* **15,** 129–136.

Nieminski, E.C. and Ongerth, J.E. (1995). Removing *Giardia* and *Cryptosporidium* by conventional treatment and direct filtration. *Jour. AWWA* **87**(9), 96–106

Nieminski, E.C. (1994). *Giardia* and *Cryptosporidium* removal through direct filtration and conventional treatment. In *Proc. AWWA Annual Conference*, 1994.

Nobel, P.J. and Medema, G.J. (1999). Removal of surrogates in treatment: pilot plant WBE. Kiwa report SWI 99.180, in Dutch.

Ongerth, J.E. and Pecoraro, J.P. (1995). Removing Cryptosporidium using multimedia filters. *Jor. AWWA* **87**(12), 83–89.

O'Shaughnessy, O.T., Barsotti, M.G., Fay, J.W., and Tighe, S.W. (1997). Evaluating particle counters. *J. Am. Wat. Assoc.* **89**(12), 60–70.

Payment, P., Trudel, M., and Plante, R. (1985). Elimination of viruses and indicator bacteria at each step of treatment during preparation of drinking water at seven water treatment plants. *App. Environ. Microbiol.* **49**(6), 1418–1428.

Payment, P. and Franco, E. (1993). *Clostridium perfringens* and somatic coliphages as indicators of the efficiency of drinking water treatment for viruses and protozoan cysts. *Appl. Environ. Microbiol.* **59**(8), 2418–2424.

Patania, N.L., Jacangelo, J.G., Cummings, L., Wilczak, A., Riley, K., and Oppenheimer, J. (1995). Optimization of Filtration for Cyst Removal. Denver, Colo: Awwa Research Foundation and American Water Works Association.

Persson, F., Långmark, J., Heinicke, G., Hedberg, T., Tobiason, J., Stenström, T., and Hermansson, M. (2005). Characterisation of barrier function of biofilters for pretreatment of drinking water with particle counts and challenge tests. *Water Res.* **39**(16), 3791–3800.

Rajala, R.L., Pulkkanen, M., Pessi, M., and Heinonen-Tanski, H. (2003). Removal of microbes from municipal wastewater effluent by rapid sand filtration and subsequent UV irradiation. *Wat. Sci. Techn.* **47**(3), 157–162.

Rao, V.Ch., Symons, J.M., Ling, A., Wang, P., Metcalf, Th.G., Hoff, J.C., and Melnick, J.L. (1988). Removal of Hepatitis A virus and rotavirus by drinking water treatment. *Jour. AWWA* 59–67.

Rice, E.W., Fox, K.R., Miltner, R.J., Lytle, D.A., and Johnson, C.H. (1996). Evaluating plant performance with endospores. *AWWA* **88**(9), 122–130.

Robeck, G.G., Clarke, N.A., and Dostal, K.A. (1962). Effectiveness of water treatment processes in virus removal. *Jour. AWWA* **54**(10), 1275–1292.

Rose, J.B., Dickson, L.J., Rarrah, S.R., and Carnahan, R.P. (1996). Removal of pathogenic and indicator micro-organisms by a full-scale water reclamation facility. *Wat. Res.* **30**(11), 2785–2797.

Schijven, J.F. and Hassanizadeh, S.M. (1999). Removal of viruses by soil passage: overview of modelling, processes and parameters. *Crit. Rev. Environ. Sci. and Technol.* **30**, 49–127.

Scott, T.M., McLaughlin, M.R.M., Harwood, V.J., Chivukula, V., Levine, A., Gennaccaro, A., Lukasik, J., Farrah, S.R., and Rose, J.B. (2003). Reduction of pathogens, indicator bacteria, and alternative indicators by waste water treatment and reclamation processes. *Wat. Sci. Techn.: Water Supply* **3**(4), 247–252.

Scott, T.M., Sabo, R.C., Lukasik, J., Boice, C., Shaw, K., Barroso-Giachetti, L., El-Shall, H., Farah, S.R., Park, C., Moudgil, B., and Koopman, B. (2002). Performance and cost-effectiveness of ferric and aluminium hydrous metal oxide coating on filter media to enhance virus removal. *KONA* **20**, 159–164.

Sriramula, N. and Chaudhurri, M. (1982). Virus removal by filtration. *Wat. Sci. Tech.* **14**, 253–256.

States, S. (1997). Protozoa in river water: sources, occurrence and treatment. *JAWWA* **89**(9), 74–83.

Swaim, P.D., Heath, M.S., Patania, N.L., Wells, W.B., and Trussell, R.R. (1996). High-rate direct filtration for *Giardia* and *Cryptosporidium* removal. In *Proc. Annual Conf. AWWA* Denver CO.

Swertfeger, J., Metz, D.H., DeMarco, J., Braghetta, A., and Jacangelo, J.G. (1997). Effect of filter media on cyst and oocyst removal. *Jour. AWWA* **91**(9), 90–100.

Tufenkji, N. (2005). Spatial distributions of retained colloidal and microbial particles in porous media: measurements, modelling and mechanisms. Thesis, Yale University, VS.

West, Th., Daniel, P., Meyerhofer, P., DeGraca, A., Leonard, S., and Gerba, C. (1994). Evaluation of *Cryptosporidium* removal through high-rate filtration. In *Proc. Annual Conf. AWWA* Denver CO.

Willis, A.T. (1957). Anaerobic bacilli in a treated water supply. *J. Appl. Bact.* **20**(1), 61–64.

Yao, K.M., Habibia, M.T., and O'Melia, C.R. (1971). 'Water and waste water filtration: concepts and applications.' *Environ. Sci. Technol.* **5**, 1105–1112.

Yates, R.S., Green, J.F., Liang, S., Merlo, R.P., and De Leon, R. (1997a). Optimizing coagulation/filtration processes for *Cryptosporidium* removal. *Int. Symp. on Waterborne Cryptosporidium*, C.R. Fricker, J.J. Clancy, P.A. Rochelle (eds), Newport Beach, US, March 1997, pp. 281–290.

Yates, R.S., Green, J.F., Liang, S., Merlo, R.P., and De Leon, R. (1997b). Optimizing filtration processes for Cryptosporidium removal. In *Proc. AWWA Water Quality Techn. Confer.*, 1997.

9.6 Slow sand filtration

Bellamy, W.D., Silverman, G.P., Hendricks, D.W., and Logsdon, G.S. (1985). Removing *Giardia* cysts with slow sand filtration. *JAWWA* **77**(2), 52–60.

Bradford, S.A., Šimůnek, J., Bettahar, M., and van Genuchten, M.Th. (2003). Modeling colloid attachment, straining and exclusion in saturated porous media. *Environ. Sci. Technol.* **37**, 2242–2250.

Bradford, S.A. and Bettahar, M. (2004). Straining, attachment and detachment of *Cryptosporidium* oocysts in saturated porous media. *J. Environ. Qual.* **34**, 469–478.

Davies, Ch., Kaucner, Ch., Altavilla, N., Ashbolt, N., Hijnen, W.A.M., Medema, G.J., Ferguson, Ch., and Deere, D. (2005). Fate and transport of surface water pathogens in watersheds. AWWARF report 2694.

Denny, S. and Broberg, P. (1992). Microbiological efficiency of water treatment: microbiological monitoring of full-scale water treatment processes. Foundation for Water Research, report FR0282.

Dullemont, Y., Visser, A., Schijven, J.F., and Hijnen, W.A.M. (2004). Elimination capacity of slow sand filtration for micro-organisms determined with spiking experiments. *H2O* **37**(13), 22–25 (in Dutch).

Dullemont, Y.F., Schijven, J.F., Hijnen, W.A.M., Colin, M., Magic, A., and Oorthuizen, W. (2006). Removal of MS2 phage, *Escherichia coli* and *Campylobacter lari* by slow sand filtration. In Recent progress in slow sand and alternative biofiltration processes ed. R. Gimbel, N.J.D. Graham and M.R. Collins, IWA publishing, London, UK.

Ellis, K.V. (1985). Slow sand filtration. In *Critical Reviews in Environmental Control*, CRC press. **15**(4), 315–354.

Cleasby, J.L., Hilmoe, D.J., and Dimitracopoulos, C.J. (1984). Slow sand and direct in-line filtration of a surface waster. *Jour. AWWA* **76**(12), 44–55.

Fogel, D., Isaac-Renton, J., Guasparini, R., Moorehead, W., and Ongerth, J. (1993). Removing *Giardia* and *Cryptosporidium* by Slow Sand Filtration. *JAWWA* **85**(11), 77–84.

Hijnen, W.A.M., Jong, R.C.M., and Medema, G.J. (2000). Dimensioning of infiltration systems for the removal of micro-organisms with column experiments. Kiwa report SWI 00-119 (in Dutch).

Hijnen, W.A.M. and Schijven, J.F. (2001). Removal of pathogenic micro-organisms by slow sand filtration. A literature review and research proposal. Kiwa report KOA 01.120.

Hijnen, W.A.M., Schijven, J.F., Bonné, P., Visser, A., and Medema, G.J. (2004). Elimination of viruses, bacteria and protozoan oocysts by slow sand filtration. *Wat. Sci. and Technol.* **50**(1), 147–154.

Hijnen, W.A.M., Brouwer-Hanzens, A.J., Charles, K.J., and Medema, G.J. (2005). Transport of MS2 phage, *Escherichia coli*, *Clostridium perfringens*, *Cryptosporidium parvum* and *Giardia intestinalis* in a gravel and a sandy soil. *Environ. Sci. & Technol.* **39**, 7860–7868.

Hijnen, W.A.M., Dullemont, Y.J., Schijven, J.F., Bosklopper, Th.G.J., and Medema, G.J. (2006a). Assessment of the capacity of slow sand filtration to eliminate *Cryptosporidium* oocysts. WQTC 5–9 November, 2006, Denver, VS.

Hijnen, W.A.M., Medema, G.J., Schijven, J.F., and Dullemont, Y.J. (2006b). Slow sand filtration fine biological safety barrier for low costs. *H2O*, 25/26, 16–17, (in Dutch).

Hijnen, W.A.M., Dullemont, Y.J., Schijven, J.F., Hanzens-Brouwer, A.J., Rosielle, M., and Medema, G.J. (2007). Removal and fate of *Cryptosporidium parvum*, *Clostridium perfringens* and small sized centric diatoms (*Stephanodiscus hantzschii*) in slow sand filters. *Water Res.* **41**, 2151–2162.

Koch, R. (1893). Wasserfiltration und Cholera. Gesammelte Werke. Bd 2 Teil 1. Georg Thieme, Leipzig, 1912.

Logan, J.L., Stevik, T.K., Siegrist, R.L., and Rønn, R.M. (2001). Transport and fate of *Cryptosporidium parvum* oocysts in intermittent sand filters. *Wat. Res.* **35**, 4359–4369.

Nobel, P.J. (1999). Effects of grazing on water quality: study at DZH, GWA and PWN. Kiwa report KOA 98.204 (in Dutch).

McConnell, L.K., Sims, R.C., and Barnett, B.B. (1984). Reovirus removal and inactivation by slow-rate sand filtration. *Appl. Environ. Microbiol.* **48**(4), 818–825.

Medema, G.J. and Theunissen, J. (1996). Elimination of viruses, *Cryptosporidium* and *Giardia* by drinking water treatment processes. RIVM report, Bilthoven, The Netherlands.

Medema, G.J., Bahar, M., and Schets, F.M. (1997). Survival of *Cryptosporidium parvum*, *Escherichia coli*, faecal enterococci and *Clostridium perfringens* in river water. *Wat. Sci. Technol.* **35**(11), 249–252.

Medema, G.J., Schets, F.M., Teunis, P.F.M., and Havelaar, A.H. (1998). Sedimentation of free and attached *Cryptosporidium* oocysts and Giardia cysts in water. *Appl. Environ. Microbiol.* **64**, 4460–4466.

Poynter, S.F.B. and Slade, J.S. (1977). The removal of viruses by slow sand filtration. *Prog. Water Tech.* **9**, 75–88.

Schijven, J.F. (2001). Virus removal from groundwater by soil passage. Thesis Technical University Delft, ISBN 90-646-4046-7.

Schijven, J.F., Hijnen, W.A.M., Collin, M., Magic, A., and Oorthuizen, W. (2008). Removal of micro-organisms by slow sand filtration. RIVM report 330204001, 2008.

Schuler, P.F., Ghosh, M.M., and Gopalan, P. (1991). Slow sand and diatomaceous earth filtration of cysts and other particulates. *Water Research* **25**(8), 995–1005.

Slade, J.S. (1978). Enteroviruses in slow sand filtered water. *J. Inst. Water Eng. Sci.* **32**, 530–536.

Timms, S., Slade, J.S., Fricker, C.R., Morris, R., Grabow, W.O.K., Botzenhart, K., and Wyn Jones, A.P. (1995). Removal of *Cryptosporidium* by Slow Sand Filtration. *Wat. Sci. and Technol.* **31**(5–6), 81–84.

Tufenkji, N., Miller, G.F., Ryan, J.N., Harvey, R.W., and Elimelech, M. (2004). Transport of *Cryptosporidium* oocysts in porous media: role of straining and physicochemical filtration. *Environ. Sci. Technol.* **38**, 5932–5938.

Visser, A., Hijnen, W.A.M., Dullemont, Y., and Medema, G.J. (2004). Slow sand filters are effective barriers for micro-organisms. *H_2O*, **37**(12), 26–28 (in Dutch).

Water Research Centre (1995). Removal of *Cryptosporidium* during water treatment. UK Water Industry Research UKWIR, report DW-06.

Wheeler, D., Bartram, J., and Lloyd, B.J. (1988). The removal of viruses by filtration through sand. In *Slow sand Filtration. Recent Developments in Water Treatment Technology*. N.J.G. Graham (ed.), Ellis Horwood Limited, Chichester.

Willis, A.T. (1956). Anaerobic bacilli in a treated water supply. *J. Appl. Bact.* **20**(1), 61–64.

Windle-Taylor, E. (1969). The removal of viruses by slow sand filtration. *Rep. Results Bac. Chem. Biol. Exam.* London Waters **44**(52), 1969–1970.

Wubbels, G. (2003). Personal communication and internal report 2000.

Yahya, M.T., Cluff, C.B., and Gerba, C.P. (1993). Virus removal by slow sand filtration and nanofiltration. *Wat. Sci. Technol.* **27**, 445–448.

9.7 UV-disinfection

Antopol, S.C. and Ellner, P.D. (1979). Susceptibility of *Legionella pneumophila* to Ultraviolet radiation. *Appl. Environ. Microbiol.* **38**(2), 347–348.

Battigelli, D.A., Sobsey, M.D., and Lobe, D.C. (1993). The inactivation of Hepatitis A virus and other model viruses by UV irradiation. *Wat. Sci. Tech.* **27**(3–4), 339–342.

Belosevic, M., Craik, S.A., Stafford, J.L., Norman, N.F., Kruithof, J.C., and Smith, D. (2001). Studies on the resistance/reactivation of *Giardia muris* cysts and *Cryptosporidium parvum* oocysts exposed to medium-pressure intraviolet radiation. *FEMS Microbiol. Letter.* **204**, 197–203.

Bernhardt, H., Hoyer, O., Hengesbach, B., Schoenen, D., Karanis, P., Moriske, H., and von Sonntag, C. (1992). Desinfektion aufbereiteter Oberflächenwässers mit UV Strahlen- erste Ergebnisse des Forschungsvorhabens. Wasser-Abwasser, **133**(12), 632–643.

Bernhardt, H. (1994). Desinfektion aufbereiteter Oberflächenwässers mit UV Strahlen. GWF Wasser-Abwasser, **135**(12), 677–689.

Bolton, J.R., Dussert, B., Bukhari, Z., Hargy, Th., and Clancy, J.L. (1998). Inactivation of *Cryptosporidium parvum* by medium-pressure Ultraviolet light in finished drinking water. In *Proceedings of the American Water Works Association Water Quality Technology Conference*, Nov. 1–5, San Diego Calif. US, 1998.

Butler, R.C., Lund, V., and Carlson, D.A. (1987). Susceptibility of *Campylobacter jejuni* and *Yersinia enterocolitica* to UV radiation. *Appl. Environ. Microbiol.* **53**(2), 375–378.

Cabaj, A., Sommer, R., and Schoenen, D. (1996). Biodosimetry: model calculations for UV water disinfection devices with regard to dose distribution. *Water Research* **30**(4), 1003–1009.

Chang, J.C.H., Ossoff, S.F., Lobe, D.C., Dorfman, M.H., Dumais, C.M., Qualls, R.G., and Johnson, J.D. (1985). UV inactivation of pathogenic and indicator micro-organisms. *Appl. Environ. Microbiol.* **49**(6), 1361–1365.

Chick, H. (1908). An investigation of the laws of disinfection. *Jour. Hygiene* **8**, 92.

Clancy, J.L., Hargy, Th.M., Marshall, M.M., and Dyksen, J.E. (1998). UV light inactivation of *Cryptosporidium* oocysts. *Journal of the American Water Works Association* **90**(9), 92–102.

Clancy, J.L., Bukhari, Z., Hargy, Th.M., Bolton, J.R., Dussert, B.W., and Marshall, M.M. (2000). Using UV to inactivate *Cryptosporidium. Journal of the American Water Works Association* **92**(9), 97–104.

Clancy, J.L., Hargy, Th.M., Battigelli, D.A., Marshall, M.M., Korich, D., and Nicholson, W.L. (2002). Susceptibility of multiple strains of *C. parvum* to UV light. Final report to American Water Works Association Research Foundation, Denver, CO.

Clancy J.L., Fallon K., Hargy Th.M., Mackey E., and Wright H. (2004). Development of non-pathogenic surrogates for large scale UV reactor validation. In *Proceedings of the American Water Works Association Water Quality Technology Conference*, Nov. 14–18, San Antonio US, 2004.

Craik, S.A., Finch, G.R., Bolton, J.R., and Belosevic, M. (2000). Inactivation of *Giardia muris* cysts using medium-pressure ultraviolet radiation in filtered drinking water. *Water Research* **34**(18), 4325–4332.

Craik, S.A., Weldon, D., Finch, G.R., Bolton, J.R., and Belosevic, M. (2001). Inactivation of *Cryptosporidium parvum* oocysts using medium- and low-pressure ultraviolet radiation. *Water Research* **35**(6), 1387–1398.

De Roda Husman, A.M., Duizer, E., Lodder, W., Pribil, W., Cabaj, A., Gehringer, P., and Sommer, R. (2003). Calicivirus inactivation by non-ionizing (UV 253.7 nm) and ionizing (gamma) radiation. In: *2nd Int. Congress on Ultraviolet Technologies*, Vienna, Austria, July 9–11, 2003.

Duizer, E., Bijkerk, P., Rockx, B., de Groot, A., Twisk, F., and Koopmans, M. (2004). Inactivation of caliciviruses. *Appl. Environ. Microbiol.* **70**(8), 4538–4543.

DVGW (1997). UV-Desinfektionsanlagen für die Trinkwasserversorgung – Anforderungen und Prüfung, Technical Standard W294. Deutscher Verein des Gas- und Wasserfaches, Bonn, Germany.

Gerba, C.P., Gramos D.M., and Nwachuku N. (2002). Comparative inactivation of enteroviruses and Adenovirus 2 by UV light. *Appl. Environ. Microbiol.* **68**(10), 5167–5169.

Haas, C.N. (1983). Estimation of risk due to low doses of micro-organisms: comparison of alternative methodologies. *Am. J. Epidemiol.* **118**(4), 573–582.

Harris, G.D., Dean Adams, V.D., Sorensen, D.L., and Curtis, M.S. (1987). Ultraviolet inactivation of selected bacteria and viruses with photoreactivation of the bacteria. *Water Research* **21**(6), 687–692.

Havelaar, A.H., Pot-Hogeboom, W.M., Koot, W., and Pot, R. (1987). F-specific bacteriophages as indicators of the disinfection efficiency of secondary effluent with ultraviolet radiation. *Ozone Sci. & Engin.* **9**, 353–368.

Havelaar, A.H., Meulemans, C.C.E., Pot-Hogeboom, W.M., and Koster, J. (1990). Inactivation of bacteriophage MS2 in wastewater effluent with monochromatic and polychromatic ultraviolet light. *Water Research* **24**(11), 1387–1393.

Henry, V., Helbronner, A., and Recklinghausen, M. (1910). Nouvelles recherches sur la sterilization de grandes quantites d'eau par les rayons ultraviolets. *Comp. Rend. Acad. Sci.* **151**, 677–680.

Hijnen, W.A.M., Veenendaal, D., van der Speld, W.M.H., Visser, A., Hoogenboezem, W., and van der Kooij, D. (2000). Enumeration of faecal indicator bacteria in large water volumes using on site membrane filtration to assess water treatment efficiency. *Water Research* **34**(5), 1659–1665.

Hijnen, W.A.M., Medema, G.J., and van der Kooij, D. (2004a). Quantitative assessment of the removal of indicator bacteria in full-scale treatment plants. *Wat. Sci. Tech.: Water Supply* **4**(2), 47–54.

Hijnen, W.A.M., van der Veer, A.J., Beerendonk, E.F., and Medema, G.J. (2004b). Increased resistance of environmental anaerobic spores to inactivation by UV. *Wat. Sci. Tech.: Water Supply* **4**(2), 54–61.

Hoyer, O. (1998). Testing performance and monitoring of UV systems for drinking water disinfection. *Wat. Supply* **16**(1/2), 424–429.

Hoyer, O. (2004). Water disinfection with UV radiation – requirements and realization. In *Proceedings European Conference UV Karlsruhe, UV radiation. Effects and Technologies*, September 22–24, 2003, Karlsruhe

Jacangelo, J.G., Loughran, P., Petrik, B., Simpson, D., and Mclloy, C. (2003). Removal of enteric viruses and selected microbial indicators by UV irradiation of secondary effluent. *Wat. Sci. Tech.* **47**(9), 193–198.

Johnson, A.M., Linden, K., Ciociola, K.M., De Leon, R., Widmer, G., and Rochelle, P.A. (2005). UV inactivation of *Cryptosporidium hominis* as measured in cell culture. *Appl. Environ. Microbiol.* **71**(5), 2800–2802.

Knudson, G.B. (1985). Photoreactivation of UV irradiated *Legionella Pneumophila* and other *Legionella* species. *Appl. Environ. Microbiol.* **49**(4), 975–980.

Kruithof, J.C., Van der Leer, R. Chr., and Hijnen, W.A.M. (1992). Practical experiences with UV disinfection in the Netherlands. *J Water SRT-Aqua* **41**(2), 88–94.

Kruithof, J.C., Kamp, P.C., Martijn, B.J., Belosevic, M., and Williams, G. (2005). UV/H_2O_2 treatment for primary disinfection and organic contaminant control at PWN's water treatment plant Andijk. In *Proceedings of Third Int. Congress on Ultraviolet Technologies*, May 24–27, Whistler, BC Canada, 2005.

Linden, K.G. and Darby, J.L. (1997). Estimating effective germicidal fluence from medium pressure UV lamps. *Jour. Envir. Engin.* 1142–1149.

Linden, K.G., Shin, G., Faubert, G., Cairns, W., and Sobsey, M.D. (2002). UV Disinfection of *Giardia lamblia* cysts in water. *Environ. Sci. Technol.* **36**, 2519–2522.

Lytke, C.D. (1971). Host-cell reactivation in mammalian calls I. survival of ultra-violet-irradiated Herpes virus in different cell-lines. *Int. Journ. of Rad. Biol.* **19**(4), 329–337.

Maier, A., Tougianidou, D., Wiedenmann, A., and Botzenhart, K. (1995). Detection of Poliovirus by cell culture and by PCR after UV disinfection. *Wat. Sci. Tech.* **31**(5–6), 141–145.

Malley, J.P., Ballester, N.A., Margolin, A.B., Linden, K.G., Mofidi, A., Bolton, J.R., Crozes, G., Laine, J.M., and Janex, M.L. (2004). Inactivation of pathogens with innovative UV technologies. American Research Foundation and American Water Works Association, 2004.

Mamane-Gravetz, H. and Linden, K.G. (2004). Impact of particle aggregated microbed on UV disinfection. In *Proceedings of the American Water Works Association Water Quality Technology Conference*, Nov. 14–18, San Antonio US, 2004.

Mamane-Gravetz, H. and Linden, K.G. (2005). Relationship between physiochemical properties, aggregation and UV inactivation of isolated environmental spores in water. *J. Appl. Microbiol.* **98**, 351–363.

Martiny, H., Schubert, T., and Rüden, H. (1990). Use of UV radiation for the disinfection of water V. Communication: microbiological investigations with bacteria of the exponential and stationary phase in cold and in warm drinking water. Zbl. Hyg. **190**, 380–394.

Maya, C., Beltrán, N., Jiménez, B., and Bonilla, P. (2003). Evaluation of the UV disinfection process in bacteria and amphizoic amoeba inactivation. *Wat. Sci. Tech.: Water Supply* **3**(4), 285–291.

Medema, G.J., Hoogenboezem, W., Van der Veer, A.J., Ketelaars, H.A.M., Hijnen, W.A.M., and Nobel, P.J. (2003). Quantitative risk assessment of *Cryptosporidium* in surface water treatment. *Wat. Sci. Tech.* **47**, 241–247.

Meng, Q.S. and Gerba, C.P. (1996). Comparative inactivation of enteric Adenoviruses, Poliovirus and coliphages by ultraviolet irradiation. *Water Research* **30**(11), 2665–2668.

Mofidi, A.A., Rochell, P.A., Chou, C.I., Mehta, H.M., Linden, K.G., and Malley, J.P. (2002). Bacterial survival after ultraviolet light disinfection: resistance, regrowth and repair. In *Proceedings of the American Water Works Association Water Quality Technology Conference*, Nov. 10–14, Seattle US, 2002.

Morita, Sh., Namikoshi, A., Hirata, T., Oguma, K., Katayama, H., Ohgaki, S., Motoyama, N., and Fujiwara, M. (2002). Efficacy of UV irradiation in inactivating *Cryptosporidium parvum* oocysts. *Appl. Environ. Microbiol.* **68**(11), 5387–5393.

Nieuwstad, Th.J. and Havelaar, A.H. (1994). The kinetics of batch ultraviolet inactivation of bacteriophage MS2 and microbiological calibration of an pilot plant. *J. Environ. Sci. Health* **A29**(9), 1993–2007.

Oguma, K., Katayama H., and Ohgaki S. (2002a). Photoreactivation of *Legionella pneumophila* after inactivation by low- or medium pressure ultraviolet lamp. *Water Research* **38**(11), 2757–2763.

Oguma, K., Katayama, H., and Ohgaki, S. (2002b). Photoreactivation of *Escherichia coli* after low- or medium-pressure UV disinfection determined by an endonuclease sensitivity site assay. *Appl. Environ. Microbiol.* **68**(12), 6029–6035.

Örmeci, B. and Linden, K.G. (2002). Comparison of UV and chlorine inactivation of particle and non-particle associated coliform. *Wat. Sci. Tech.: Water Supply* **2**(5–6), 403–410.

Österreichisches Normungsinstitut (1999). Anlagen zur Desinfektion von Wasser mittels Ultraviolett-strahlen, Anforderungen und Prüfung. Önorm M 5873-1, Wien.

Oppenheimer, J.A., Hoagland, J.E., Laine, J.-M., Jacangelo, J.G., and Bhamrah, A. (1993). Microbial inactivation and characterization of toxicity and by-products occurring in reclaimed wastewater disinfected with UV radiation. Planning, design and operation of effluent disinfection systems, 13–24. Whippany, N.J.: the Water Environment Federation.

Oppenheimer, J.A., Jacangelo, J.G., Laîné, J.M., and Hoagland, J.E. (1995). Equivalency testing of ultraviolet disinfection for wastewater reclamation. In *Proceedings of the American Water Works Association Water Quality Technology Conference*, November 12–16, New Orleans US, 1995.

Pirnie, M., Linden, K.G., and Malley, J.P. (2003). Ultraviolet disinfection guidance manual. EPA 815-D-03-007, June 2003.

Qian, S.S., Donnelly M., Pirnie, D.C., Messner M., Linden, K.G., and Cotton, C. (2004). Ultraviolet light inactivation of protozoa in drinking water: a Bayesian meta-analysis. *Water Research* **38**(2), 317–326.

Qualls, R.G. and Johnson, J.D. (1983a). Bioassay and fluence measurement in UV disinfection. *Appl. Environ. Microbiol.* **45**, 872–877.

Qualls, R.G., Flynn, M.P., and Johnson, J.D. (1983b). The role of suspended particles in ultraviolet disinfection. *J. Water Pollut. Control Fed.* **55**(10), 1280–1285.

Regli S., Rose J.B., Haas C.N., and Gerba C.P. (1991). Modelling the risk from *Giardia* and viruses in drinking water. *Journal of American Water Works Association* **83**, 76–84.

Rochelle, P.A., Mofidi, A.A., Marshall, M.M., Upton, S.J., Montelone, B., Woods, K., and Di Giovanni, G. (2004). An investigation of UV disinfection and repair in *Cryptosporidium parvum.* Report of the American Water Works Association AWWA, 2004.

Rontó, G., Gáspár, S., and Bérces, A. (1992). Phage T7 in biological UV dose measurement. *J. Photochem. Photobiol. B Biol.* **12**, 285–294.

Schoenen, D., Zemke, V., and Kolch, A. (1991). The influence of reflection of the UV Rays on the Disinfection of Drinking Water. Zbl. Hyg. **191**, 396–405.

Schoenen, D. Kolch, A., Gebel, J., and Hoyer, O. (1995). UV-Desinfektion von Trinkwasser, Grundlagen und Anforderungen für die praktische Anwendung. *DVGW-Schriftenreihe Wasser* **86**, 303–350.

Severin, B.F., Suidan, M.T., and Engelbrecht, R.S. (1983). Effects of temperature on ultraviolet light disinfection. *Environ. Sci. Technol.* **17**, 717–721.

Shin, G., Linden, K.G., Arrowood, M.J., and Sobsey, M.D. (2001). Low-pressure UV inactivation and DNA repair potential of *Cryptosporidium parvum* oocysts. *Appl. environ. Microbiol.* **67**(7), 3029–3032.

Shin, G., Bohrerova, Z., Linden, K.G., and Faubert, G. (2005). DNA repair of UV-irradiated *Giardia lamblia* cysts detected by both infectivity and molecular biological assays. In *proceedings of third Int. Congress on Ultraviolet Technologies*, May 24–27, Whistler, BC Canada, 2005.

Simpson, D., Jacangelo, J., Loughran, P., and McIlroy, C. (2003). Investigation of potential surrogate organisms and public health risk in UV irradiated secondary effluent. *Wat. Sci. Tech.* **47**(9), 37–43.

Smeets, P.W.M.H., van der Helm, A.W.C., Dullemont, Y., Rietveld, L., van Dijk, H., and Medema, G.J. (2004). *E. coli* inactivation by Ozone Under Bench-scale and Full-scale Hydraulic Conditions. Submitted to Water Research, November 2004.

Smeets, P.W.M.H., Dullemont, Y., and Medema, G.J. (2005). *E. coli* as surrogate for *Campylobacter* inactivation at bench-scale and full-scale and high ozone resistance of environmental *E. coli* and *Campylobacter.* In *Proceedings of the 17th Conference of International Ozone Association*, August 22–26, 2005.

Sommer, R., Weber, G., Cabaj, A., Wekerle, J., Keck, G., and Schauberger, G. (1989). UV inactivation of micro-organisms in water. *Zbl. Hyg.* **189**, 214–224.

Sommer, R., Haider, Th., Cabaj, A., Heidenreich, E., and Kundi, M. (1996a). Increased inactivation of *Sacharomyces cerevisiae* by protraction of UV irradiation. *Appl. Environ. Microbiol.* **62**(6), 1977–1983.

Sommer, R., Cabaj, A., and Haider, Th. (1996b). Microbiocidal effect of reflected UV radiation in devices for water disinfection. *Wat. Sci. Tech.* **34**(7–8), 173–177.

Sommer, R., Cabaj, A., Pribil, W., and Haider, T. (1997). Influence of lamp intensity and water transmittance on the UV disinfection of water. *Wat. Sci. Tech.* **35**(11–12), 113–118.

Sommer, R., Haider, Th., Cabaj, A., Pribil, W., and Lhotsky, M. (1998). Time fluence reciprocity in UV disinfection of water. *Wat. Sci. Tech.* **38**(12), 145–150.

Sommer, R., Cabaj, A., Sandu, T., and Lhotsky, M. (1999). Measurement of UV radiation using suspensions of micro-organisms. *J. Photochem. Photobiol. B: Biol.* **53**, 1–6.

Sommer, R., Lhotsky, M., Haider, Th., and Cabaj, A. (2000a). UV inactivation, liquid-holding recovery, and photoreaction of *Escherichia coli* OH157 and other pathogenic *Escherichia coli* strains in water. *Jour. Food Prot.* **63**(8), 1015–1020.

Sommer, R., Cabaj, A., Pribil, W., Haider, Th., and Hirschmann, G. (2000b). Differences between calculated and biodosimetrically measured fluences in UV plants for drinking water disinfection – Practical experiences with the Austrian National Standard M 5873-1. In: *Proc. IOA, European-African-Asian-Australasian Group, Wasser Berlin*, October 23–26, 2000.

Sommer, R., Pribil, W., Appelt, S., Gehringer, P., Eschweiler, H., Leth, H., Cabaj, A., and Haider, Th. (2001). Inactivation of bacteriophages in water by means of non-ionizing (UV-253.7 nm) and ionizing (gamma) radiation: a comparative approach. *Water Research* **35**(13), 3109–3116.

Teunis, P.F.M., Medema, G.J., Kruidenier, L., and Havelaar, A.H. (1997). Assessment of the risk of infection by *Cryptosporidium* or *Giardia* in drinking water from a surface water source. *Water Research* **31**(16), 1333–1346.

Thompson, S.S., Jackson, J.L., Suva-Castillo, M., Yanko, W.A., El Jack, Z., Kuo, J., Chen, Ch., Williams, F.P., and Schnurr, D.P. (2003). Detection of infectious human Adenoviruses in tertiary-treated and ultraviolet-disinfected wastewater. *Wat. Environ. Res.* **75**(2), 163–170.

Thurston-Enriquez, J.A., Haas, Ch.N., Jacangelo, J., Riley, K., and Gerba, Ch.P. (2003). Inactivation of Feline Calicivirus and Adenovirus Type 40 by UV radiation. *Appl. Environ. Microbiol.* **69**(1), 577–582.

von Sonntag, C., Kolch, A., Gebel, J., Oguma, K., and Sommer, R. (2004). The photochemical basis of UV disinfection. In *Proc. European Conference UV Karlsruhe, UV radiation. Effects and Technologies*, September 22–24, 2003, Karlsruhe.

Watson, H.E. (1908). A note on variation of the rate of disinfection with change in the concentration of the disinfectant. *Jour. Hygiene* **8**, 536.

Watercare (2002). Pilot plant investigations, surrogate study, results and recommendations. Disinfection review group report Water Care Services Ltd., New Zealand, 2002.

Wilson, B.R., Roessler, P.F., van Dellen, E., Abbaszadegan, M., and Gerba, C.P. (1992). Coliphage MS2 as a UV water disinfection efficacy test surrogate for bacterial and viral pathogens. In *Proceedings of American Water Works Association Water Quality Technology Conference*, Nov. 15–19, Toronto Canada, 1992.

Wiedenmann, A., Fischer, B., Straub, U., Wang, C.H., Flehmig, B., and Schoenen, D. (1993). Disinfection of Hepatitis A virus and MS2 coliphage in water by ultraviolet irradiation: comparison of UV susceptibility. *Wat. Sci. Tech.* **27**(3–4), 335–338.

Wolfe, R.L. (1990). Ultraviolet disinfection of potable water. *Environ. Sci. Technol.* **24**(6), 768–773.

Zimmer, J.L. and Slawton, R.M. (2002). Potential repair of *Escherichia coli* DNA following exposure to UV radiation from both medium- and low-pressure UV sources used in drinking water treatment. *Appl. Environ. Microbiol.* **68**(7), 3293–3299.

Zimmer, J.L., Slawson, R.M., and Huck, P.M. (2003). Inactivation and potential repair of *Cryptosporidium parvum* following low- and medium-pressure ultraviolet irradiation. *Water Research* **37**(14), 3517–3523.

IWA Publishing's authorised EU representative for General Product Safety Regulations is Diane D'Arras, 15 rue Duret, 75116 Paris, France, e-mail: safety@iwap.co.uk.

Printed and bound by CPI Group (UK) Ltd, Croydon, CR0 4YY
12/05/2026
02108863-0018